Post-Genome Informatics

Post-Genome Informatics

Minoru Kanehisa

Institute for Chemical Research, Kyoto University, Japan

OXFORD
UNIVERSITY PRESS

OXFORD

UNIVERSITY PRESS

Great Clarendon Street, Oxford OX2 6DP

Oxford University Press is a department of the University of Oxford.
It furthers the University's objective of excellence in research, scholarship,
and education by publishing worldwide in

Oxford New York

Athens Auckland Bankok Bogotá Buenos Aires Kolkata
Cape Town Chennai Dares Salaam Delhi Florence Hong Kong Istanbul
Karachi Kuala Lumpur Madrid Melbourne Mexico City Mumbai
Nairobi Paris São Paulo Shanghai Taipei Tokyo Toronto Warsaw

with associated companies in Berlin Ibadan

Oxford is a registered trade mark of Oxford University Press
in the UK and in certain other countries

Published in the United States
by Oxford University Press Inc., New York

A catalogue record for this book is available from the British Library

Library of Congress Cataloging in Publication Data
Kanehisa, Minoru
Post-genome informatics / Minoru Kanehisa
p. cm
1. Genomes—Data processing. 2. Medical informatics. I. Title.
QH447 .K365 1999   572.8'6'0285—dc21   99-049584

ISBN 0 19 850326 1 (Pbk)

Typeset by J&L Composition Ltd, Filey, North Yorkshire
Printed in Great Britain
on acid-free paper by
Biddles Ltd, Guildford & King's Lynn

Preface

The Human Genome Project was initiated in the late 1980s as the result of technological developments in molecular biology and with the expectation of biomedical benefits. The project's goal was to determine the entire sequence of three billion nucleotides in the human genome and to identify and understand a whole repertoire of human genes. At the same time, genome projects were conceived and undertaken for a number of organisms from bacteria to higher eukaryotes. Both the public and private sectors are spending unprecedented amounts of resources in order to quickly decipher the genomes and to claim discovery of the information. However, the determination of the complete genome sequence is not the end of the story. It is actually the beginning of 'post-genome informatics', especially in view of the fact that the biological function cannot be inferred from the sequence information alone for roughly one half of the genes in every genome that has been sequenced.

Conceptually, whole genome sequencing represents an ultimate form of reductionism in molecular biology. It is hoped that complex processes of life can be explained by simple principles of genes. In experimental reality, DNA sequencing requires drastic reductions from higher to lower dimension—to destroy the cell and to extract the DNA molecules. We do not question how much information is lost in these procedures, but simply accept the common wisdom that the genome, or the entire set of DNA molecules, contains all the necessary information to make up the cell. Post-genome informatics is then considered as an attempt at synthesis from lower to higher dimension, whereby the functioning biological system of a cell is reconstructed from the entire complement of genes.

The genome projects have transformed biology in many ways, but the most impressive outcome is the emergence of computational biology, also known as bioinformatics. It is no longer possible to make advances in biology without integration of informatics technologies and experimental technologies. Here we like to distinguish between genome informatics and post-genome informatics. Genome informatics was born in order to cope with the vast amount of data generated by the genome projects. Its primary role is therefore to support experimental projects. In contrast, post-genome informatics, as we define here, represents a synthesis of biological knowledge from genomic information toward understanding basic principles of life, as well as for practical purposes in biomedical applications. Post-genome informatics has to be coupled with systematic experiments in functional genomics using DNA chip and other technologies. However, the coupling is the other way around—informatics plays more dominant roles of making predictions and designing experiments.

This book is an introduction to bioinformatics, an interdisciplinary science encompassing biology, computer science, and physics. In fact the major motivation for writing this book is to provide conceptual links between different disciplines, which often share common ideas and principles. The content is in part a translation of my book in Japanese *Invitation to genome informatics* (Kyoritsu Shuppan, Tokyo, 1996), which originated from my lecture notes on theoretical molecular biology for undergraduate students in the Faculty of Science, Kyoto University. The first chapter is a concise introduction to molecular biology and the Human Genome Project. The second and third chapters provide an overall picture of both database and computational issues in bioinformatics. They are written for basic understanding of underlying concepts rather than for acquiring the superficial skills of using specific databases or computational tools. Because most algorithmic details are deliberately left out in order to cover a wide range of computational methods, it is recommended that the reader consult the references in the Appendix where necessary.

The last chapter, which has been specially written for this English edition, is the essence of post-genome informatics. It introduces the emerging field of network analysis for uncovering systemic functional information of biological organisms from genomic information. KEGG (Kyoto Encyclopedia of Genes and Genomes) at *www.genome.ad.jp/kegg/* is a practical implementation of databases and computational tools for network analysis. It is our attempt to actually perform synthesis of biological systems for all the genomes that have been sequenced. Since the field of network analysis is likely to evolve rapidly in the near future, KEGG should be considered as an updated version of the last chapter.

The very concept of post-genome informatics grew out of my involvement in the Japanese Human Genome Program. I have been supported by the Ministry of Education, Science, Sports and Culture since 1991 as princpal investigator of the Genome Informatics Project. This book is the result of active collaborations and stimulating discussions with the many friends and colleagues in this project. I am grateful to Chigusa Ogawa, Hiroko Ishida, Saeko Adachi, and Toshi Nakatani for their work on the drawings and to Stephanie Marton for checking the text of the English edition. The support of the Daido Life Foundation is also appreciated.

With a diverse range of Internet resources publicly available, it is not difficult for anyone interested to start studying post-genome informatics. I hope this book will help students and researchers in different disciplines to understand the philosophy of synthesis in post-genome informatics, which is actually the antithesis of the extreme specialization found in current scientific disciplines. The study of post-genome informatics may eventually lead to a grand synthesis—a grand unification of the laws in physics and biology.

Minoru Kanehisa

Kyoto, Japan
May 1999

Contents

3 Sequence analysis of nucleic acids and proteins

1

Blueprint of life

Gene and genome

Life is a complex system for information storage and processing. Information is transmitted 'vertically' from cell to cell and from generation to generation, while at the same time information is expressed 'horizontally' within a cell in the ontogenesis of an individual organism. The information transmission from parent to offspring must have been recognized vaguely as heredity since the early days of human history. However, it was Gregor Mendel's experiments on the garden pea, performed in the mid-1800s but not recognized until 1900, that provided a first glimpse of its mechanism. A hereditary unit, later called a gene, was found to determine a particular characteristic, or a trait, of an organism. The Mendelian law of inheritance has established, first of all, that genes are inherited more or less independently in the vertical flow of information transmission and, secondly, that there is a set of competing genes (alleles) so that what is inherited (genotype) is not necessarily what is observed (phenotype) as the result of the horizontal flow of information expression.

Biology is the science of life that aims at understanding both functional and structural aspects of living organisms. In the latter half of the nineteenth century, great advances were made not only in genetics, a branch of biology analysing functional appearances of heredity, but also in cell biology based on microscopic observation of cellular structures. Most importantly, it was discovered that the chromosome in the nucleus of a cell contains the hereditary material. The entire set of genes in the chromosome, or more precisely in the haploid chromosome, was later named the genome. In view of the dual flow of information in life, the genome can be defined as the structural and functional unit of the information transmission and the gene as the structural and functional unit of the information expression (Table 1.1).

The disciplines of genetics and cell biology that emerged in the nineteenth century are the roots of modern biology, especially the golden era of molecular biology in the latter half of the twentieth century. The elaboration of

Table 1.1. Genome and gene

Entity	Definition	Molecular mechanism
Genome	Unit of information transmission	DNA replication
Gene	Unit of information expression	Transcription to RNA Translation to protein

experimental technologies in molecular biology has established the molecular basis of genes and genomes, uncovered the structure–function relationships of biological macromolecules in a diverse range of cell processes, and ultimately led to the conception of the Human Genome Project, a project working towards deciphering the blueprint of life. Let us quickly overview these developments.

DNA and protein

The chromosome is a molecular complex made from deoxyribonucleic acids (DNAs) and proteins. Originally, the protein was suspected to be the genetic material, but by the mid-1900s it became apparent that the DNA contained the information transmitted and the protein was synthesized within the cell. In 1953 James Watson and Francis Crick proposed the double helix model for the DNA structure. The model was constructed from the X-ray diffraction data for the DNA fibres, which had been obtained by Rosalind Franklin and Maurice Wilkins, together with the experimental observation that in any DNA the composition of adenine (A) plus thymine (T) was equal to the composition of guanine (G) plus cytosine (C). In essence, a DNA molecule contains two chains, where each chain is a linear polymer consisting of the repeating units of the four nucleotides, A, T, G, and C (Tables 1.2 and 1.3). The structural complementarity and the resulting hydrogen bonding between A and T and between G and C stabilize the assembly of the two chains into the double helix structure (Fig. 1.1). The genetic informa-

Table 1.2. Nucleic acid and protein

Macromolecule		Backbone	Repeating unit	Length	Role
Nucleic acid	DNA	Phosphodiester bonds	Deoxyribonucleotides (A, C, G, T)	10^3–10^8	Genome
	RNA	Phosphodiester bonds	Ribonucleotides (A, C, G, U)	10^3–10^5 10^3–10^4 10^2–10^3	Genome Messenger Gene product
Protein		Peptide bonds	Amino acids (A, C, D, E, F, G, H, I, K, L, M, N, P, Q, R, S, T, V, W, Y)	10^2 – 10^3	Gene product

Table 1.3. Nucleotide codes

A	Adenine	W	Weak (A or T)
G	Guanine	S	Strong (G or C)
C	Cytosine	M	Amino (A or C)
T	Thymine	K	Keto (G or T)
U	Uracil	B	Not A (G or C or T)
R	Purine (A or G)	H	Not G (A or C or T)
Y	Pyrimidine (C or T)	D	Not C (A or G or T)
N	Any nucleotide	V	Not T (A or G or C)

(a)   (b)

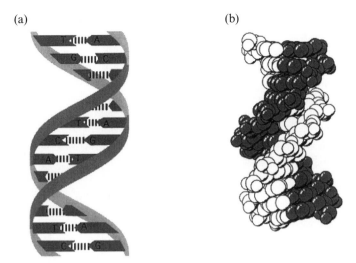

Fig. 1.1. DNA double helix. (a) A schematic diagram of the Watson–Crick model. (b) The three-dimensional structure of a synthetic DNA, 'CGCGAATTCGCG' (PDB:7BNA).

tion is encoded in the sequence of nucleotides that forms the polymer, and once the nucleotide sequence of one chain is given, the other sequence is automatically determined by the complementarity. Thus, the DNA double helix model has a profound implication for the molecular mechanism of heredity. The transmission

Table 1.4. Amino acid codes

Ala	A	Alanine
Arg	R	Arginine
Asn	N	Asparagine
Asp	D	Aspartic acid
Cys	C	Cysteine
Gln	Q	Glutamine
Glu	E	Glutamic acid
Gly	G	Glycine
His	H	Histidine
Ile	I	Isoleucine
Leu	L	Leucine
Lys	K	Lysine
Met	M	Methionine
Phe	F	Phenylalanine
Pro	P	Proline
Ser	S	Serine
Thr	T	Threonine
Trp	W	Tryptophan
Tyr	Y	Tyrosine
Val	V	Valine
Asx	B	Asn or Asp
Glx	Z	Gln or Glu
Sec	U	Selenocysteine
Unk	X	Unknown

of genetic information is realized by the replication of DNA molecules (Table 1.1) in which the complementary chains guarantee correct copying of the information.

The protein is a linear polymer of the 20 different kinds of amino acids, which are linked by peptide bonds (Tables 1.2 and 1.4). The three-dimensional structure of a protein that results from the folding of the polypeptide chain is far more complex than the double helical DNA structure. This complexity reflects the variety and specificity of protein functions. The amino acid sequence, or the primary structure of a protein, was first determined for insulin by Frederick Sanger in 1953. The three-dimensional (3D) structure, or the tertiary structure, was first elucidated for myoglobin by John Kendrew in 1960 using the X-ray diffraction of protein crystals. The X-ray diffraction was refined by Max Perutz and became known as the isomorphous replacement method. The tertiary structure of myoglobin is an assembly of six α-helices as shown in Fig. 1.2. The model structure of an α-helix was originally proposed for homopolymers of single amino acids by Linus Pauling in 1951, who also predicted the existence of α-helices in proteins. As more 3D structures were resolved general principles emerged for the functioning of proteins. It was again the structural complementarity, as in the case of the

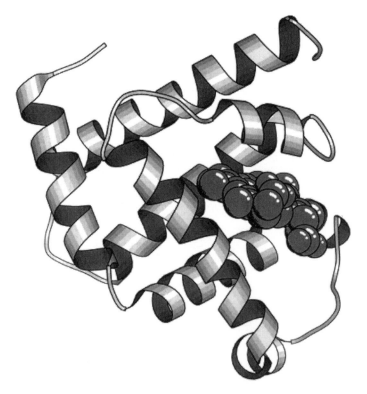

Fig. 1.2. The three-dimensional structure of sperm whale myoglobin (PDB:1MBN).

DNA double helix, for example, that enabled enzymes to recognize and react on substrates. The enzyme–substrate complementarity is like the matching of a key-hole with a specific key, but the analogy should not be taken in a strict sense since molecular structures are flexible and inducible. It is the chemical complementarity in the structural framework that ultimately determines the interaction of two molecules.

Central dogma

DNAs and proteins are the biological macromolecules that play key roles in the living cell. Both are linear polymers of repeating units (Table 1.2). The genetic information is encoded and stored in the sequence of four types of nucleotides, and DNA replication is the molecular mechanism for genetic information transmission. In contrast, the actual functioning of the cell is mostly performed by proteins. Thus, the conversion of a DNA sequence with four possible 'letters' into a protein sequence with twenty possible 'letters' is required. This is a crucial step in the genetic information expression. The molecular mechanism of this conversion is the translation where a triplet of nucleotides, or a codon, is translated into an amino acid. The translation table of 64 codons into 20 amino acids is specified by the genetic code shown in Table 1.5. Originally the genetic code was considered to be universal among all species, but now a number of variations are known, as shown in Table 1.6.

Table 1.5. Standard genetic code

1st Position	2nd position								3rd Position
	U		C		A		G		
U	UUU	Phe	UCU	Ser	UAU	Tyr	UGU	Cys	U
	UUC	Phe	UCC	Ser	UAC	Tyr	UGC	Cys	C
	UUA	Leu	UCA	Ser	UAA	*Stop*	UGA	*Stop*	A
	UUG	Leu	UCG	Ser	UAG	*Stop*	UGG	Trp	G
C	CUU	Leu	CCU	Pro	CAU	His	CGU	Arg	U
	CUC	Leu	CCC	Pro	CAC	His	CGC	Arg	C
	CUA	Leu	CCA	Pro	CAA	Gln	CGA	Arg	A
	CUG	Leu	CCG	Pro	CAG	Gln	CGG	Arg	G
A	AUU	Ile	ACU	Thr	AAU	Asn	AGU	Ser	U
	AUC	Ile	ACC	Thr	AAC	Asn	AGC	Ser	C
	AUA	Ile	ACA	Thr	AAA	Lys	AGA	Arg	A
	AUG	Met, Start	ACG	Thr	AAG	Lys	AGG	Arg	G
G	GUU	Val	GCU	Ala	GAU	Asp	GGU	Gly	U
	GUC	Val	GCC	Ala	GAC	Asp	GGC	Gly	C
	GUA	Val	GCA	Ala	GAA	Glu	GGA	Gly	A
	GUG	Val	GCG	Ala	GAG	Glu	GGG	Gly	G

Table 1.6. Variation of genetic codes

	T1	T2	T3	T4	T5	T6	T9	T10	T12	T13	T14	T15
CUU	Leu	–	Thr	–	–	–	–	–	–	–	–	–
CUC	Leu	–	Thr	–	–	–	–	–	–	–	–	–
CUA	Leu	–	Thr	–	–	–	–	–	–	–	–	–
CUG	Leu	–	Thr	–	–	–	–	–	Ser	–	–	–
AUU	Ile	–	–	–	–	–	–	–	–	–	–	–
AUC	Ile	–	–	–	–	–	–	–	–	–	–	–
AUA	Ile	Met	Met	–	Met	–	–	–	–	Met	–	–
AUG	Met	–	–	–	–	–	–	–	–	–	–	–
UAU	Tyr	–	–	–	–	–	–	–	–	–	–	–
UAC	Tyr	–	–	–	–	–	–	–	–	–	–	–
UAA	Stop	–	–	–	–	Gln	–	–	–	–	Tyr	–
UAG	Stop	–	–	–	–	Gln	–	–	–	–	–	Gln
AAU	Asn	–	–	–	–	–	–	–	–	–	–	–
AAC	Asn	–	–	–	–	–	–	–	–	–	–	–
AAA	Lys	–	–	–	–	–	Asn	–	–	–	Asn	–
AAG	Lys	–	–	–	–	–	–	–	–	–	–	–
UGU	Cys	–	–	–	–	–	–	–	–	–	–	–
UGC	Cys	–	–	–	–	–	–	–	–	–	–	–
UGA	Stop	Trp	Trp	Trp	Trp	–	Trp	Cys	–	Trp	Trp	–
UGG	Trp	–	–	–	–	–	–	–	–	–	–	–
AGU	Ser	–	–	–	–	–	–	–	–	–	–	–
AGC	Ser	–	–	–	–	–	–	–	–	–	–	–
AGA	Arg	Stop	–	–	Ser	–	Ser	–	–	Gly	Ser	–
AGG	Arg	Stop	–	–	Ser	–	Ser	–	–	Gly	Ser	–

T1, Standard code; T2, vertebrate mitochondrial code; T3, yeast mitochondrial code; T4, mould, proto-zoan, and coelenterate mitochondrial code and mycoplasma and spiroplasma code; T5, invertebrate mitochondrial code; T6, ciliate, dasycladacean and hexamita nuclear code; T9, echinoderm mitochondrial code; T10, euplotid nuclear code; T12, alternative yeast nuclear code; T13, ascidian mitochondrial code; T14, flatworm mitochondrial code; T15, blepharisma nuclear code.

The conversion of DNA information into protein information is not direct; DNA is first transcribed to RNA (ribonucleic acid) which then is translated to protein. This particular type of RNA is called messenger RNA (mRNA), since other types of RNAs also exist including transfer RNA (tRNA) and ribosomal RNA (rRNA). The RNA is another linear macromolecule that is closely related to DNA; the only differences are that the sugar backbone is made of ribose rather than deoxyribose and that thymine (T) is substituted by uracil (U) (Tables 1.2 and 1.3). In a symbolic representation, the transcription is to change only the letter T in the DNA sequence to the letter U to obtain the RNA sequence. Thus, there is a unidirectional flow of information expression from DNA to mRNA to protein. This flow, together with the flow of information transmission from DNA to DNA, forms the central dogma of molecular biology (Fig. 1.3) as first enunciated by Francis Crick in 1958.

(Present World)

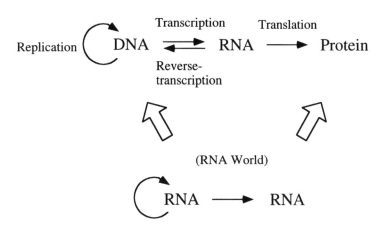

Fig. 1.3. The central dogma and its evolution.

However, it soon became known that in a special virus called a retrovirus there appeared to be an opposite flow of information—from RNA to DNA. This reverse-transcription was predicted by Howard Temin and the enzyme responsible for the reaction, reverse-transcriptase, was discovered in 1970. Retroviruses are an intriguing class of viruses that include many cancer viruses and the AIDS virus, and they store the genetic information in genomic RNA rather than DNA. The genetic information transmission thus requires the reverse-transcription from RNA to DNA before the DNA replication.

In 1977 another surprising discovery was made. The mRNA transcript was not entirely used for the following step of translation. The protein coding gene was found to be divided into pieces, which is called the exon/intron structure. There is an additional step of mRNA processing where introns are removed and exons are spliced to form a mature mRNA molecule. This mRNA splicing is prevalent in higher species, and occurs in the nucleus after transcription from genomic DNA. It is then followed by the nuclear export of mature mRNA and the translation of protein on the ribosome. RNA splicing is also known to exist in tRNA genes and rRNA genes in certain species. Furthermore, in contrast to RNA splicing that is an operation of nucleotide deletions, another type of processing called RNA editing that involves insertions and replacements of specific nucleotides has been discovered in a limited group of species. The post-transcriptional RNA processing does not contradict the central dogma in terms of the direction of information flow, but it certainly has caused modifications on the complexity of RNA information processing.

RNA world

In almost all of the species that presently exist on Earth, the information of the genome is stored and transmitted in DNA and the information of the gene is expressed to the functional molecule of protein. Is RNA's main role, as the central dogma implies, simply to be a messenger for the more important molecules of DNA and protein? Quite the contrary. It is widely believed that life started with RNA and there was an RNA world when neither DNA nor protein yet existed. RNA must have been playing a dual role both as the genome in the information transmission and as the functional molecule in the information expression. In fact, RNA can be a genome because there still exist viruses with RNA genomes. RNA can be a functional molecule, first because tRNAs and rRNAs are functional molecules expressed from genes, but more importantly because a new class of RNAs with catalytic activities was discovered by Thomas Cech in 1981. The discovery of this ribozyme transformed the idea that the catalysis of chemical reactions was exclusively performed by protein enzymes. The repertoire of catalysis by ribozymes has since been increasing, partly from additional discovery of natural ribozymes but mostly from the design and selection of artificial ribozymes by *in-vitro* molecular evolution experiments.

If the RNA world existed, then the central dogma would have evolved as shown in Fig. 1.3. In terms of the stability of information storage and the fidelity of copying, DNA is a more favourable molecule than RNA. In terms of the variety and specificity of catalytic functions, protein is a more favourable molecule than RNA. These could have been the reasons that DNA and protein, with their separate roles, respectively, for information storage (template) and information processing (catalyst), replaced the dual roles of RNA. The bizarre phenomena of RNA splicing and RNA editing could be remnants of RNA processing in the RNA world to produce catalytic molecules from templates in genomic RNA. The fact that ribose rather than deoxyribose is often seen in biologically active chemicals, such as ATP, cyclic AMP, NAD, and coenzyme A, could also be considered a remnant of the RNA world. The variety of genetic codes (Table 1.6) could suggest that the central dogma evolved somewhat independently in different species.

Then, how could the transition from the RNA world to the DNA/protein world happen? The transition from RNA to DNA is conceivable because they are highly similar molecules and they can be used in an identical manner for information storage. In contrast, RNA and protein are quite different molecules. However, what needs to be preserved here is not the capacity of information storage but the capacity of information processing. Therefore, as long as the catalytic activity was preserved, the switch from RNA to protein could happen in the evolution of life, like switching from cathode ray tube to transistor in the early days of computer development. Since the catalytic activity is highly related to the structural and chemical complementarity of interacting molecules, the keyhole–key relation of interface structures must have been preserved during the transition of the two worlds.

Cell

As we have seen, a most fundamental aspect of life is information storage and processing. The information is stored in molecules and the information is processed by molecules. The information processing to copy the storage is called information transmission, and the information processing to make catalysts from templates in the storage is called information expression. While the central dogma neatly summarizes the molecules that are involved in the two types of information processing, it does not state when and where the processing should take place. In fact, life would not exist without the cell, which provides the field for space-dependent and time-dependent processing.

The concept that all biological organisms are made up of structural and functional units, or cells, was established by the early nineteenth century. The cell contains smaller, or subcellular, units such as the nucleus, which stores genetic information. Biological organisms have traditionally been classified into two domains by the cellular architecture—eukaryotes that are made up of cells with a nucleus and prokaryotes that are made up of cells without a nucleus. Note, however, that biological organisms can also be divided into three domains of life by the

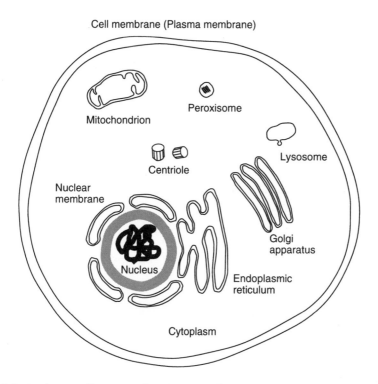

Fig. 1.4. A schematic illustration of a eukaryotic cell.

molecular architecture, with prokaryotes being split into bacteria and archaea. Figure 1.4 is a schematic drawing of a eukaryotic cell, which consists of cytoplasm, nucleus, endoplasmic reticulum, Golgi apparatus, mitochondria, and other organelles. The cell and each of the organelles are surrounded by special membranes that allow only limited sets of molecules to be transported. The cellular architecture is critical in providing spatial environments for the information processing by molecules.

In addition to the space-dependent environment, the cell provides the time-dependent environment. A cell divides into two cells according to an internal clock or in response to external stimuli. There are two types of cell divisions, meiosis and mitosis, that correspond to the vertical flow of information transmission and the horizontal flow of information expression, respectively, as illustrated in Fig. 1.5. A reproductive cell, or a gamete, is produced by meiosis, which then becomes a germ cell to start development of a new organism by mitosis. Therefore, in all organisms on Earth in the present or in the past there is a cellular continuity that leads to the origin of life. This is called the germ cell line.

In the traditional view, the genome is a blueprint of life that determines all aspects of organism development given a proper environment. This view implicitly assumes that the genome is a commander of creating life, so that everything else becomes an environment. However, it must be emphasized that what is transmitted in the vertical flow of life is the entire cell, not just the genome or the nucleus.

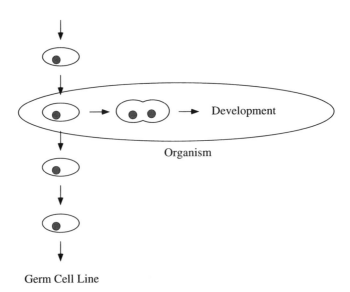

Fig. 1.5. The dual flow of information in life: the vertical flow by genetic information transmission and the horizontal flow by the genetic information expression.

What if the commander resides at a different place in the cell? We will come back to this point later in this chapter.

Technological developments in molecular biology

The history of molecular biology has been a history of technological developments, especially for determining the primary and tertiary structures of the biological macromolecules—DNAs, RNAs, and proteins. New experimental technologies for structure determinations have greatly advanced all aspects of life sciences, for the knowledge of molecular structures provides clues to understanding molecular functions. This is the principle of the structure–function relationship. Figure 1.6 is a selected list of landmark events in the history of structure determinations in molecular biology, some of which have already been mentioned.

The Watson–Crick model of the DNA double helix was based on low-resolution X-ray fibre diffraction, and it was only in the late 1970s that high-resolution X-ray diffraction became available for single crystals of short synthetic DNAs. With the atomic resolution the DNA structure was found to be far more polymorphic than previously thought, as shown in Fig. 1.7. The Watson–Crick DNA is called B-form, this is a right-handed helix with about 10 base pairs per pitch, it contains a major groove and a minor groove, and consists of base pairs that are oriented roughly perpendicular to the axis of the helix. Under low water content DNA takes a more skewed conformation called A-form, and under high salt concentration and depending on the sequence a left-handed helix called Z-form is also observed. Furthermore, even within the B-form double helix, the local structures exhibit significant sequence-dependent fluctuations from the average Watson–Crick structure.

The primary structure of an RNA molecule was first determined in 1965 by Robert Holley for alanyl-transfer RNA, which was a decade before the DNA sequencing method became available. The tRNA sequence contained self-complementary portions that would form Watson–Crick base pairing and, as shown in Fig. 1.8, the cloverleaf model of tRNA secondary structure was presented. The model was later confirmed by the 3D structure of a phenylalanyl-transfer RNA obtained by X-ray diffraction in 1973, which also revealed the tertiary base pairing to stabilize the L-shaped structure. The RNA double helix is known to be in the A-form. As in proteins, the specific tertiary structure is the basis of the specific function in RNAs.

The discovery of restriction enzymes in the 1960s led to the development of recombinant DNA technology, including DNA sequencing methods established around 1975. The type II restriction enzyme recognizes a specific pattern in the double helical DNA and cuts both chains around the recognition site. For example, EcoRI isolated from *Escherichia coli* recognizes the pattern 'GAATTC' and introduces a cut between 'G' and 'A' as shown below.

	Technology development	**Structure determination**
1950	49 Edman degradation	
		51 α-helix model
	54 Isomorphous replacement	53 DNA double helix model Insulin primary structure
1960		60 Myoglobin tertiary structure
	62 Restriction enzyme	
		65 tRNA^{Ala} primary structure
1970		
	72 DNA cloning	73 tRNA^{Phe} tertiary structure
	75 DNA sequencing	
		77 ϕX174 complete genome
1980		79 Z-DNA by single crystal diffraction
	84 Pulse field gel electrophoresis 85 Polymerase chain reaction	86 Protein structure by 2D NMR
	87 YAC vector	88 Human Genome Project
1990		
	93 DNA chip	
		95 *H. influenzae* complete genome
2000		

Fig. 1.6. History of structure determinations for nucleic acids and proteins.

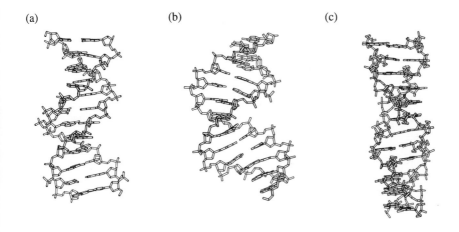

(a)   (b)   (c)

Fig. 1.7. Polymorphic DNA tertiary structures: (a) B-type DNA (PDB:7BNA); (b) A-type DNA (PDB:140D); (c) Z-type DNA (PDB:2ZNA).

$$\downarrow$$
$$5' - G\ A\ A\ T\ T\ C - 3'$$
$$3' - C\ T\ T\ A\ A\ G - 5'$$
$$\uparrow$$

Because the two chains of DNA are assembled in the anti-parallel direction, and because the recognition pattern is self-complementary, i.e. containing a dyad symmetry, the two cuts will leave single-stranded ends, called sticky ends, on both strands that can then be used to connect other sequences of DNA. Thus, the restriction enzyme is a type of selective cutter, creating recognizable sequences that allow for specific pasting.

Another important operation on DNA sequences is to make a large number of copies. Molecular cloning involves inserting a specific DNA segment into a vector that is to be amplified in bacterial or yeast cells, while PCR (polymerase chain reaction) is a cell-free amplification system using a special enzyme. Both the Sanger method and the Maxam–Gilbert method of DNA sequencing are based on the idea of separating DNA segments by electrophoresis according to their lengths. The successive improvements of DNA sequencing methods, developments of new cloning vectors, and advances in other technologies, such as that used for separating large DNA segments, eventually made it possible to decode human chromosomes.

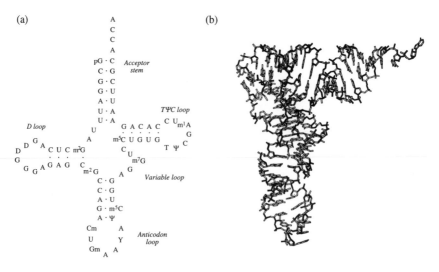

Fig. 1.8. Transfer RNA. (a) The primary sequence and the secondary structure of yeast alanyl-transfer RNA. (b) The tertiary structure of yeast phenylalanyl-transfer RNA (PDB:1TRA).

Human Genome Project

In 1977 a tiny virus genome, $\phi x174$, consisting of just 5000 nucleotides and 11 genes was sequenced by the emerging technology of DNA sequence determination (Fig. 1.6). After two decades of technology developments the first complete genome sequence of a free-living organism, *Haemophilus influenzae*, was determined in 1995. The bacterial genome determination, consisting of 1.8 million nucleotides and 1700 genes, has been followed by an explosion of determinations of complete genome sequences due to the coordinated efforts for a number of organisms from bacteria to higher eukaryotes. The ultimate goal of the Human Genome Project is to uncover three billion nucleotides and 100 000 genes in the human chromosomes, and to make use of the information for new discoveries and applications in biomedical sciences. The project was initiated in the late 1980s and is expected to be finished, at least in terms of the sequence determination, early in the 2000s.

Figure 1.9 is a comparison of the genome sizes together with the history of genome sequence determinations of different organisms. Roughly speaking, viruses have genome sizes in the range of 10^3 to 10^5 nucleotides, while free-living organisms have 10^6 to 10^9 nucleotides. In general, higher organisms have larger genome sizes, but the human genome is not the largest. Perhaps a better indication of higher organisms is the total size of transcribed regions containing codes for functional RNA and protein molecules. The genomes of bacteria and archaea are in the range of 10^6 nucleotides and contain genes, on the average, every 1000

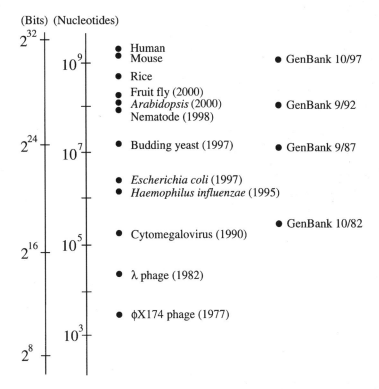

Fig. 1.9. The information content of various species in terms of the number of nucleotides in the genome. The complete genome sequences were determined in the years as designated. The increase of the GenBank nucleotide sequence database (see *Nucleotide sequence database* in Section 2.1) is also shown together with the release dates.

nucleotides. The genome of a single-cell eukaryote, *Saccharomyces cerevisiae*, is about 10^7 nucleotides and contains genes roughly every 2000 nucleotides. The genome of a simple multicellular organism, *Caenorhabditis elegans*, is about 10^8 nucleotides and contains genes roughly every 5000 nucleotides. Thus, there is a definite tendency that the increasing genome size is accompanied by an increasing fraction of non-coding regions.

Since the initiation of the Human Genome Project computational biology, also known as bioinformatics, has become an integral part of molecular biology. A primary role of bioinformatics has been to organize and manage the data and to assist experimental projects with new informatics technologies. At the same time, the informatics infrastructure has been established to make a diverse range of databases and computational resources widely available over the Internet. More importantly, the next major task of bioinformatics is to make biological sense out of the vast amount of data—not only the sequence data that are generated by the high-throughput DNA sequencing technology, but also the expression and mutation data that will be generated by the new DNA chip technologies.

The biological function is not known for roughly half of the genes in every genome that has been sequenced. This is partly due to the lack of appropriate informatics technologies to interpret sequence information, but it is largely due to the fact that the genome itself does not tell much about the process of genetic information expression. A new level of systematic experiments is required to obtain an overall picture of when, where, and how genes are expressed. The study of functional genomics includes the analysis of gene expression profiles at the mRNA and protein levels and the analysis of polymorphism or mutation patterns in the genome. DNA chips are high-density arrays of DNA segments used for hybridization experiments, and are especially suited to obtain such expression and mutation profiles. Functional clues can also be obtained by observing phenotypic consequences of disrupting or overexpressing specific genes. Such genetic manipulations are of course impossible in humans, but they can be substituted by ample clinical data once disorders are linked to specific genes.

The nucleus of a human somatic cell contains 46 chromosomes: 22 pairs of autosomes and two sex chromosomes (Fig. 1.10). Autosomes are numbered according to their approximate sizes. Sex chromosomes are denoted by X and Y, XX being for female and XY for male. Because of the single X chromosome, a mutation for X-linked recessive disorders can affect males as illustrated in Fig. 1.11(a). The other types of genetic disorders are autosomal dominant and autosomal recessive. The genes that are responsible, either entirely or partly, for a number of diseases have been identified by the linkage analysis, which is illustrated in Fig. 1.11(b). It is based on the principle that due to the crossing-over between homologous pairs of chromosomes during meiosis, the distance between two chromosomal sites can be measured by the frequency that the two sites are transmitted together to offspring. Once the human genome is covered by a set of markers whose locations are known, the linkage between the morbidity and one of the markers will tell the location of a candidate disease gene. A marker can be anything that is polymorphic among individuals and whose inheritance patterns can be detected. Different eye colours and other traits may be used as markers, but polymorphisms in DNA sequences, such as single nucleotide polymorphisms (SNPs), are far more abundant and are generally used today.

Reductionism in biology

Biology has been an empirical discipline dominated by the progress of experimental technologies to obtain different types of data. Technologies in molecular biology were especially powerful and, as a consequence, reductionism was the mainstream of biology in the latter half of the twentieth century. In the reductionistic approach, starting from a specific functional aspect of the biological system (organism), the building blocks (genes and proteins) that are responsible for the

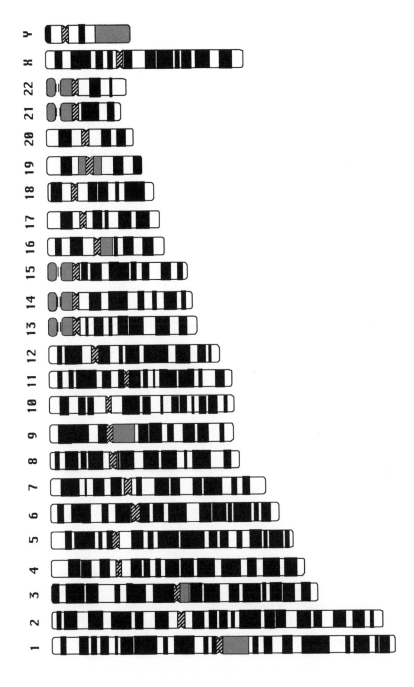

Fig. 1.10. A schematic illustration of human chromosomes.

(a)

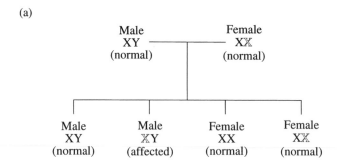

(b)

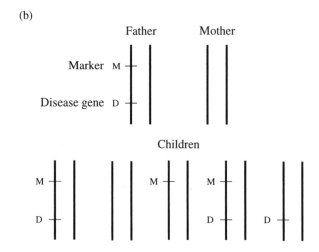

Fig. 1.11. (a) X-linked recessive disorder. The inheritance pattern is shown for a recessive gene on the chromosome X, designated in outline. (b) Linkage analysis. The relative distance between the disease gene (D) and a marker (M) can be estimated from the frequency that both D and M are transmitted together to offspring.

function are searched for and characterized (Fig. 1.12(a)). In contrast, the complete set of genes and gene products that has become available by the genome project is the starting point of an alternative approach, which may be called a synthetic approach, toward understanding how genes and molecules are networked to form a biological system. The synthetic approach is an informatics based approach that requires technologies to process massive data, to clarify general principles, and to compute systemic behaviours from the building blocks.

In physics and chemistry how matter is organized from elementary particles and how chemical compounds are organized from chemical elements are relatively well-known general principles (Fig. 1.12(b)). In biology such principles are largely unknown. In fact, we do not yet know whether the information in the genome is suf-

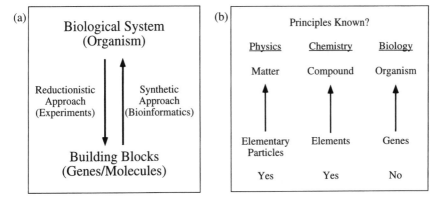

Fig. 1.12. (a) Reductionistic and synthetic approaches in biology. (b) Basic principles in physics, chemistry, and biology.

ficient to build an entire biological system. In the standard model of elementary particle physics, there are two categories of elementary particles: the constituents of matter and the transmitters of force. Duality of particle and wave is inherent in matter. In other words, the information on the building blocks alone is not sufficient; the information on their interactions is absolutely essential. The genome certainly contains the information on the building blocks, but it is premature to assume that the genome also contains the information on how to connect or relate the building blocks.

In addition to the concept of interaction, another important concept is the level of abstraction. As illustrated in Fig. 1.13, cysteine is a network of carbon, nitrogen, oxygen, hydrogen, and sulfur atoms at the atomic level, but it is abstracted to the letter C at the molecular level where a protein is represented by a one-dimensional sequence of twenty letters (amino acids). In the next molecular network level the protein is abstracted to a symbol, Ras, and the wiring among different symbols (proteins) becomes a major concern as in the case of the Ras signal transduction pathway. Post-genome informatics, as we define here, is a challenging branch of bioinformatics that aims at understanding biology at the molecular network level. It uses various types of data generated by the genome projects—not only the sequence data but also the other types of data such as gene expression profiles and genome diversity profiles. Eventually it will become necessary to consider still higher levels to understand, for example, brain functions, but the molecular network level is a starting point and may turn out to be general enough to describe phenomena at higher levels as well.

Grand challenges in post-genome informatics

The protein folding problem has been a grand challenge in computational molecular biology. The challenge is to predict the native 3D structure of a protein from

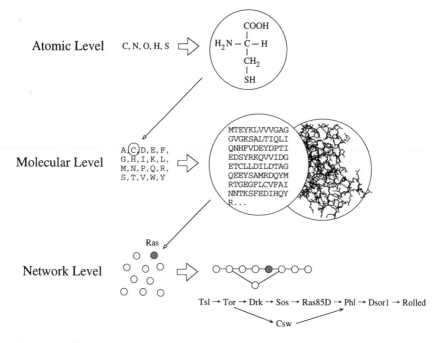

Fig. 1.13. The concept of the level of abstraction.

its amino acid sequence. According to the experiments first performed by Christian Anfinsen in the 1950s, protein unfolding (denaturation) and refolding (renaturation) are thermodynamically reversible processes. By changing the environmental conditions, such as temperature, pressure, and solvent conditions, the protein unfolds and loses its activity, but once the environment is returned to the physiological condition the protein folds up spontaneously to its native structure and regains its activity. Thus, protein folding is apparently thermodynamically determined and the amino acid sequence appears to contain all the necessary information to make up the native 3D structure.

Anfinsen's thermodynamic principle is well established for small globular proteins under *in-vitro* experimental conditions, but protein folding in the living cell is a more complex and dynamic process involving a number of other molecules, such as chaperones. The cellular environment is probably not a smooth thermodynamic environment, but rather a reflection of specific interactions with various molecules. Therefore, an alternative view of the protein folding problem is that each protein is only a part of the whole system and that the amino acid sequence does not necessarily contain all the information to fold up the protein molecule. It is not unreasonable to expect that the protein folding problem cannot be solved for the majority of proteins in nature without considering the network of specific molecular interactions.

In the era of whole genome sequencing, we are faced with a new grand challenge problem, which may be called the organism reconstruction problem. Given a complete genome sequence, the problem is to reconstruct in a computer the functioning system of a biological organism. This involves the prediction of all wiring (interactions) among the building blocks (genes and molecules). Here again, a traditional view is that the genome is a blueprint of life containing all necessary information that would make up a biological organism. A clone can be made by replacing the nucleus of a cell (the localized area containing all genetic information); however, what is transmitted from parent to offspring is not just the nucleus but the entire cell. Thus, an alternative view can also be taken where the genome is only a part of the whole network of interacting molecules in the cell. The genome is not the headquarters of instructions; rather it is simply a warehouse of parts. The headquarters reside in the network itself which has an intrinsic capacity, or a programme, to undergo development and to reproduce gamete cells.

Whichever view one takes, it is impossible in practice to make sense fully out of the complete genome sequence data without additional information, especially information on molecular wiring. Although none of the attempts to actually compute the native 3D structure from the amino acid sequence information alone have been successful, the protein folding problem can be solved in practice whenever the matching reference data exist in the known set of 3D structures. Similarly, the organism reconstruction problem can be approached by making use of knowledge of reference organisms. Toward this end, first, it is extremely important to computerize the current knowledge on molecular wiring that has resulted from all relevant experiments in genetics, biochemistry, and molecular and cellular biology. Second, new experiments have to be designed to systematically detect molecular wiring. Figure 1.14 illustrates functional genomics experiments where perturbation–response relations of the living cell are used, together with the complete genome sequence and incomplete biological knowledge, to uncover the underlying wiring diagrams. With this new level of information, the organism reconstruction problem may one day be solved.

Genetic and chemical blueprints of life

Life is a manifestation of both genetic and chemical information. Although the biological macromolecules, DNAs, RNAs, and proteins, play dominant roles in the current working of life, small chemical substances and metal ions must have played important roles in the origin and evolution of life. Their roles must still be significant now, for the environment in which life exists consists of those substances. As repeatedly emphasized in this chapter, the information in the genome is not sufficient to make up life. An essence of life is the dynamics of the information flow, both vertical (information transmission) and horizontal (information expression), which is an ordered sequence and collection of chemical reactions. In fact, the reactions of the biological macromolecules that underlie, for example, transcription

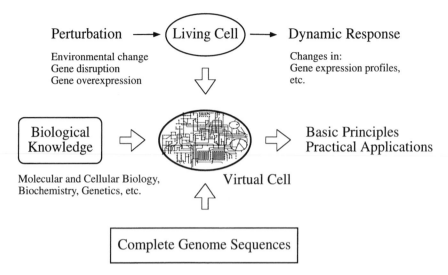

Fig. 1.14. Virtual cell—computer representation and simulation of a biological organism.

or signal transduction, are no different from chemical reactions of small molecules except that they are highly complex yet ordered. We call this ordered network of complex reactions the chemical blueprint of life. The genome certainly is an information storage centre for the entire set of genes; the specification of molecular templates is the genetic blueprint of life. However, the genome does not contain the entire programme of reactions. Perhaps it contains only a fraction of the programme, for, after all, the DNA molecules are part of the reaction network too.

Extraordinary achievements made by molecular biology in the twentieth century have had profound impacts on the modern notions of life science: for one is reductionism and another is genetic determinism. Ironically, the ultimate form of reductionism, the sequencing of the entire complement of the genome, has raised scepticism. The reason why identical twins are not really identical does not necessarily contradict the principle of genetic determinism because it can be due to the uncertainty of the environment. A question is raised as to the very concept of what the genetic material is and what the environment is. The genetic material, or the information transmitted, is the genome according to current wisdom, but we emphasize that it should be considered as the entire germ cell to include the programme of reactions. In this view, the genome is a dynamic entity that may be modified in the network of reactions. The Mendelian law of inheritance implicitly assumes that the information in the genome is static; discrete units of heredity (genes) pass unchanged from generation to generation. Although evidence of moving genes across species was presented by Barbara McClintock as early as the 1950s in her experiments on maize, this concept was not accepted for a long time and it is still considered as a process that occurs in special cases in limited species rather than as a general principle. On the other hand, life would not have evolved and would not continue to evolve without exchange of information across species.

This leads to a hierarchical view of how to distinguish between the genetic material and the environment. The genome is only a part of what makes up an individual organism, but it probably contains the most information within the network of interacting molecules. The complete genome sequence is the best approximation of identity. At a higher level of abstraction, an individual organism becomes a part of a whole. Conceptually, the entire universe of life is considered as a network of interacting individual organisms, which is approximated by a distribution profile in a hypothetical space of all possible genomic sequences. Species would correspond to local clusters in this space and the grouping of species would be seen as larger clusters by smoothing the space. Furthermore, the time dependence and the environment dependence of this space would represent both birth-to-death of individuals and birth-to-death of species. The principles that underlie the dynamics of the universe of life would be consistent with the principles that govern the dynamics of Earth in the physical world, although the current laws of physics and chemistry are probably insufficient. The ultimate objective of post-genome informatics is therefore to unite life and matter and to establish a grand-unification theory of the physical and biological worlds.

2

Molecular biology databases

2.1 Historical background

Evolution of molecular biology databases

The current state of biology is arguably comparable to seventeenth century physics in terms of the stage of formulating the science. Biology is still an empirical discipline with little formulation of basic principles that can be used in a deductive way to predict various biological phenomena. The technologies of genetics and cell biology, started in the nineteenth century, and of molecular biology, started in the twentieth century, have finally made it possible to gather fundamental observables in biology. It is expected that empirical rules similar to Kepler's laws and basic principles similar to Newton's theory will come along. This is where the computerized databases play critical roles as the basis of our expanding knowledge in biology.

In any scientific discipline, the database is first developed for computerizing information from published materials. A bibliographic database is a secondary publication that contains the abstract of a primary publication together with its author, title, publisher, and other citation information, as well as indexed keywords and classification codes. The purpose of the bibliographic database is information retrieval. Once the information is found the user would either go to the library and read the printed journal or, as is becoming more prevalent now, view the on-line version of the full-text article. In any event, literature databases (Tables 2.1 and 2.2) like the bibliographic databases and on-line journals are intended for humans to read and understand.

The result reported in a scientific paper is often based on experimental data, which may or may not be printed in a journal article. A factual database is a collection of experimental data that are associated with published articles and that can be used for computerized analysis. In molecular biology the primary factual data-

Table 2.1. Evolution of molecular biology databases

	Database category	Data content	Examples
1	Literature database	Bibliographic citations On-line journals	MEDLINE (1971)
2	Factual database	Nucleic acid sequences Amino acid sequences 3D molecular structures	GenBank (1982), EMBL (1982), DDBJ (1984) PIR (1968), PRF (1979), SWISS-PROT (1986) PDB (1971), CSD (1965)
3	Knowledge base	Motif libraries Molecular classifications Biochemical pathways	PROSITE (1988) SCOP (1994) KEGG (1995)

Table 2.2. The addresses for the major databases

Database	Organization	Address
MEDLINE	National Library of Medicine	www.nlm.nih.gov
GenBank	National Center for Biotechnology Information	www.ncbi.nlm.nih.gov
EMBL	European Bioinformatics Institute	www.ebi.ac.uk
DDBJ	National Institute of Genetics, Japan	www.ddbj.nig.ac.jp
SWISS-PROT	Swiss Institute of Bioinformatics	www.expasy.ch
PIR	National Biomedical Research Foundation	www-nbrf.georgetown.edu
PRF	Protein Research Foundation, Japan	www.prf.or.jp
PDB	Research Collaboratory for Structural Bioinformatics	www.rcsb.org
CSD	Cambridge Crystallographic Data Centre	www.ccdc.cam.ac.uk

bases are for nucleic acid sequences, protein sequences, and 3D molecular structures (Tables 2.1 and 2.2). Because of the large data volume, such as that of DNA sequences and 3D atomic coordinates, the journal article reports a summary of the result and the complete data can only be found in the databases. Thus, the nucleic acid sequence databases and the 3D molecular structure database have become a mode of scientific publication and, consequently, the data organization is not necessarily appropriate from a biological point of view. These databases are essentially repositories of all published data with standardization of data formats but without enough standardization of data contents.

This leads to the next category of databases, which may broadly be called knowledge bases (Table 2.1). A knowledge base is intended for inference rather than simple retrieval. Knowledge is different from data in that new knowledge can be generated from stored knowledge. Just as a pile of published articles does not represent organized knowledge, a collection of factual data does not itself represent biological knowledge. This is especially because the factual databases in molecular biology contain information on molecular structures, i.e. sequences and 3D structures, which cannot be automatically related to biological functions. The main theme of this chapter is to discuss the attempts that are being made to organize a third generation of databases that focus on representing and utilizing the knowledge on biological functions (Table 2.3). However, let us first review the history of the literature databases and the factual databases that are the backbone databases in molecular biology (Table 2.2).

Bibliographic databases

The history of the bibliographic databases dates back to the era before the computer was invented. *Chemical abstracts* (CA) has been published since 1907 by the Chemical Abstracts Service (CAS) of the American Chemical Society, and *Biological abstracts* (BA) since 1926 by the non-profit organization BIOSIS. Both CA and BA started as secondary printed publications containing the abstracts and bibliographic citations for the primary scientific literature, including journals, books, conference proceedings, and patents. They are the largest collections, CA for

Table 2.3. New generation of molecular biology databases

Information	Database	Address
Compounds and reactions	LIGAND	www.genome.ad.jp/dbget/ligand.html
	AAindex	www.genome.ad.jp/dbget/aaindex.html
Protein families and	PROSITE	www.expasy.ch/prosite
sequence motifs	Blocks	www.blocks.fhcrc.org/
	PRINTS	www.bioinf.man.ac.uk/dbbrowser/PRINTS/
	Pfam	www.sanger.ac.uk/Pfam/, pfam.wustl.edu/
	ProDom	protein.toulouse.inra.fr/prodom.html
3D fold classifications	SCOP	scop.mrc-lmb.cam.ac.uk/scop/
	CATH	www.biochem.ucl.ac.uk/bsm/cath/
Orthologous genes	COG	www.ncbi.nlm.nih.gov/COG/
	KEGG	www.genome.ad.jp/kegg/
Biochemical pathways	KEGG	www.genome.ad.jp/kegg/
	WIT	wit.mcs.anl.gov/WIT2/
	EcoCyc	ecocyc.PangeaSystems.com/ecocyc/
	UM-BBD	umbbd.ahc.umn.edu/
Genome diversity	NCBI Taxonomy	www.ncbi.nlm.nih.gov/Taxonomy/
	OMIM	www.ncbi.nlm.nih.gov/Omim/

chemical literature and BA for biological and medical literature though overlaps do exist, and growing at a rate of over a half million records every year. Since 1965 CAS has also been producing the factual database, CAS Registry, for a comprehensive collection of chemical substances, which is growing at a rate of over one million records per year. The CAS Registry number is a unique identifier that can be used to link many different resources.

The National Library of Medicine (NLM) of the National Institutes of Health (NIH) has been indexing biomedical literature to produce *Index medicus*, and its on-line counterpart MEDLINE has been available since 1971. The number of covered sources is somewhat smaller than CA or BA, but MEDLINE has become the premier bibliographic database for molecular biology since the National Center for Biotechnology Information (NCBI) started its operation in 1988. NCBI has transformed MEDLINE in the following ways. First, MEDLINE is fully linked with the factual databases of DNA sequences, protein sequences, 3D molecular structures, and others. Second, MEDLINE is linked to many of the publishers who provide full-text journals on-line. Third, MEDLINE has become freely available over the Internet, which is in contrast to most of the bibliographic databases and chemical databases that require paid subscription.

Amino acid sequence databases

Since the protein sequencing method became available in the late 1950s, comparative studies with interests in molecular evolution were initiated by individual researchers. Among them Margaret Dayhoff at the National Biomedical Research Foundation (NBRF) in Washington, DC was most active in collecting all available amino acid sequences. She published the collection in the printed series *Atlas of protein sequences and structures* from 1968 to 1978. The concept of superfamilies was developed from the collection, and the mutation data matrix, called MDM78 or PAM250, was compiled from the observed frequency of amino acid mutations (see *Amino acid indices* and *Protein families and sequence motifs* in Section 2.3). Around 1980, the time of competition for the establishment of a national DNA database in the US, the collection was computerized and called the NBRF protein sequence database. The PIR (Protein Information Resource) database evolved from the NBRF database and was established in 1984 with support from the NIH. Since 1988 PIR is collaborating with the Munich Information Center for Protein Sequences (MIPS) in Germany and the Japanese International Protein Sequence Database (JIPID) in Japan to produce the PIR–International Protein Sequence Database.

The Protein Research Foundation (PRF) in Osaka, Japan has been producing a printed publication *Peptide information* since 1975, which contains bibliographic citations for reports on peptides and proteins. The computerized database is called LITDB and its companion SEQDB is a protein sequence database that was initiated in 1979 as a by-product of their efforts. Both SEQDB and LITDB were integrated and cross-referenced with each other from the beginning, which was long before the GenBank–MEDLINE links were introduced by NCBI. Although SEQDB contains minimal information other than the sequence data, it remains a unique resource that still covers the field of protein sequencing. All the other protein sequence databases rely on nucleic acid sequence databases, for the great majority of protein sequences are now determined by DNA sequencing.

NBRF and PRF started their databases from contrasting points of views. NBRF treated the amino acid sequence as a biological entity. They spent major efforts on annotation, superfamily classification, and cleaning duplicate entries, which unfortunately left them unable to keep up with the rate of data increase. In contrast, PRF treated the amino acid sequence as part of bibliographic information just like the abstract. They could cope with the increased data flow, but the usefulness of their database was limited. All the sequence databases, both protein and nucleic acid, have had the problem of how to balance biology and bibliography with an ever increasing amount of data.

SWISS-PROT was created in 1986 at the University of Geneva as another protein sequence database, but it soon became the best in terms of the data quality. It is a value-added database containing rich annotation with good curation and numerous links to other databases. SWISS-PROT has been collaborating with the European Molecular Biology Laboratory (EMBL) since 1987 and the translation of the EMBL nucleotide sequence database (TREMBL) has been used to supplement SWISS-PROT. Over the years, the backlog of translated entries that need be processed has become so large that SWISS-PROT is facing the same old problem of how to balance the quality and quantity. SWISS-PROT is now operated jointly by the Swiss Institute of Bioinformatics (SIB) and the European Bioinformatics Institute (EBI) of EMBL.

Three-dimensional structure databases

Since protein crystallography became available, the need for computerizing relevant data in a public repository became immediately apparent. The Protein Data Bank (PDB) was established in 1971 at the Bookhaven National Laboratory (BNL) as an archive of experimentally determined three-dimensional structures of biological macromolecules, originally by X-ray crystallography but now by NMR as well. In 1999 the PDB operation was moved to the Research Collaboratory for Structural Bioinformatics (RCSB). Despite its naming, PDB covers structural data for proteins, RNAs, short DNAs, carbohydrates, molecular complexes, and also viruses. PDB is basically a collection of text files, each representing a structure entry deposited by the authors. An entry contains atomic coordinates, bibliographic citations, primary and secondary structure information, together with crystallographic structure factors and NMR experimental data. Because PDB is an archive with divergent sets of data and because 3D structural data are far more complex than sequence data, there have been many attempts to reorganize and refine the data in PDB to produce secondary databases, some of which will be covered later in this chapter (see *Classification of protein 3D structures* in Section 2.3).

The Cambridge Structural Database (CSD) contains three-dimensional structure data for organic and metal organic compounds, some of which are biologically important. CSD is compiled by the Cambridge Crystallographic Data Centre (CCDC), which was established in 1965 at the University of Cambridge and which became a non-profit institution in 1989. CSD contains three-dimensional molecular coordinate data determined by X-ray and neutron diffraction methods together with associated bibliographic, chemical, and crystallographic data. It is increasing at a rate of about 15 000 structures per year.

Nucleotide sequence databases

By the end of the 1970s it was apparent that the emerging DNA sequencing technology would transform biology with a flood of sequence data. National and international efforts to establish DNA data banks were initiated around 1979 in both the United States and Europe. Subsequently, the GenBank database at the Los Alamos National Laboratory and the EMBL database at the European Molecular Biology Laboratory were officially initiated in 1982. GenBank and EMBL have been in collaboration since then, and the DNA Data Bank of Japan (DDBJ) joined in 1984. In 1992 the role of GenBank was assumed by NCBI, and in 1994 the EMBL operation was also moved to the newly established outstation EBI. GenBank, EMBL, and DDBJ form the International Nucleotide Sequence Database Collaboration. All receive data by authors' direct submissions and exchange data on a daily basis. The contents of GenBank and DDBJ are almost identical. EMBL enters additional links to other databases although the sequence data itself is virtually the same.

Figure 2.1 shows the growth of the sequence and 3D structure databases in the past two decades. The initial increase of GenBank and EMBL was caused by the decision to enter data without much annotation to clear the backlog. They then increased steadily with a doubling time of about 20 months until 1995, when massive EST sequencing and whole genome sequencing shortened the doubling time to about 14 months. PDB suddenly started growing at a higher rate around 1992 probably because of the wide availability of protein crystallography. The figure also reveals that the number of entries in the protein sequence databases is not keeping up with the increase of entries seen in the nucleotide sequence databases.

Among the major molecular biology databases, the nucleotide sequence databases were the last to appear. They were thus in the best position to adopt new computer technologies and to prepare for the imminent data explosion, but they actually had the most turbulent history. Initially it was the responsibility of the database producers to find journal articles, manually enter the data, and annotate sequences. Then, the power of searching sequence databases became apparent due to the discoveries of unexpected sequence similarities, such as between viral oncogenes and cellular genes. Because there was such a huge backlog of unprocessed data the pressure started to mount to just keep up the data flow, leaving the annotation to authors. A coordinated effort was also initiated to convince journal publishers to require database submission as a precondition of publication. Thus, the mode of data entry has changed from manual input by the database producers to direct submission from the authors. Consequently, the work shared among the three databases has changed from a division based on journal title to a division based on geographic location of the author. More importantly, the consequence of this change was the transformation of the database itself from an annotated database of biological information to a comprehensive repository of bibliographic information. Although the annotation is still the major concern in SWISS-PROT and, to some extent, in PIR, it is necessary to reconsider how to develop databases of biological information.

Flat file format

The current databases for nucleic acid sequences, protein sequences, and 3D molecular structures rely on sophisticated database management systems, but they all started as flat file databases. In the flat file format a database is considered a sequential collection of entries, which may be stored in a single text file or multiple text files. When the amount of data was not large, primitive database management systems that had been developed by individual groups allowed the databases to be successfully organized and maintained in the flat file format. The flat file format was also instrumental in promoting the wide use of molecular biology databases, since programs could easily be written to handle and utilize the data. Although the major database developers subsequently started using more efficient and reliable database management systems, such as the relational database

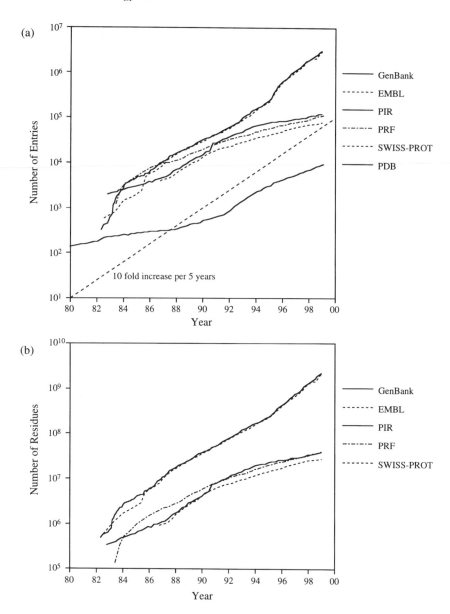

Fig. 2.1. Growth of the sequence and 3D structure databases in terms of (a) the number of entries and (b) the number of residues.

management system, the flat file format is still widely in use as a mode of database distribution.

Figure 2.2 shows examples of database entries in GenBank and SWISS-PROT in the flat file format. An entry consists of multiple records, each of which is identified by a key word such as LOCUS and DEFINITION in GenBank or by a two-letter code such as ID and DT in SWISS-PROT. The entry name that is the unique identifier of an entry is given in the LOCUS line or in the ID line. For historical reasons the sequence databases contain another set of identifiers called the accession numbers. One or more accession numbers appear in the ACCESSION line or in the AC line. The first one is unique and is called the primary accession number. Traditionally, the entry names have been used to somehow represent or classify the data content, for example, by containing codes to specify organism names (DROME represents *Drosophila melanogaster* in SWISS-PROT in Fig. 2.2(b)). In contrast, the accession numbers are just combinations of letters and numbers. They were implemented to relate to the release history including the original publication. When two entries are merged the entry name may be the same as either of the two previous ones, but a new accession number is assigned as the primary accession number, with the previous two numbers remaining as secondary accession numbers.

The content of a sequence database entry generally consists of three parts. The first part contains the nomenclature and bibliographic information. The primary sequence data forms the third part. In between these two parts is a features table that contains biological annotations of sequence features in a computer readable form (see the FEATURES record in GenBank or the FT lines in SWISS-PROT in Fig. 2.2). This information is indispensable in analysing, for example, sequence patterns around promoter regions or splice junctions. Each database defines specific keys to allow retrieval of such features and corresponding sequence data. A database in molecular biology often starts with personal research interests where the quality of biological annotation, as well as the quantity of data collection, is often critical. As can easily be imagined, the annotation of biological features requires continuous efforts for improvement of individual entries. Now that the nucleic acid sequence databases have grown too large to perform comprehensive, consistent, and up-to-date annotation across all entries, and that the features tables are created separately by individual authors, the interpretation of the annotated data should be performed with caution.

Genome databases

The primary role of molecular biology databases was to disseminate information of published or finished works. Thus, the major databases described above (Table 2.2) are open to the general public. As the genome projects started to unfold, databases assumed a new role of assisting on-going experimental works. The database is now considered part of the informatics technology that is essential for managing

```
(a) LOCUS       DRODPPC      4001 bp    mRNA           INV      15-MAR-1990
    DEFINITION  D.melanogaster decapentaplegic gene complex (DPP-C), complete cds.
    ACCESSION   M30116
    NID         g157291
    KEYWORDS    .
    SOURCE      D.melanogaster, cDNA to mRNA.
      ORGANISM  Drosophila melanogaster
                Eukaryotae; mitochondrial eukaryotes; Metazoa; Arthropoda;
                Tracheata; Insecta; Pterygota; Diptera; Brachycera; Muscomorpha;
                Ephydroidea; Drosophilidae; Drosophila.
    REFERENCE   1  (bases 1 to 4001)
      AUTHORS   Padgett,R.W., St Johnston,R.D. and Gelbart,W.M.
      TITLE     A transcript from a Drosophila pattern gene predicts a protein
                homologous to the transforming growth factor-beta family
      JOURNAL   Nature 325, 81-84 (1987)
      MEDLINE   87090408
    COMMENT     The initiation codon could be at either 1188-1190 or 1587-1589.
    FEATURES              Location/Qualifiers
         source          1..4001
                         /organism="Drosophila melanogaster"
                         /db_xref="taxon:7227"
         mRNA            <1..3918
                         /gene="dpp"
                         /note="decapentaplegic protein mRNA"
                         /db_xref="FlyBase:FBgn0000490"
         gene            1..4001
                         /note="decapentaplegic"
                         /gene="dpp"
                         /allele=""
                         /db_xref="FlyBase:FBgn0000490"
         CDS             1188..2954
                         /gene="dpp"
                         /note="decapentaplegic protein (1188 could be 1587)"
                         /codon_start=1
                         /db_xref="FlyBase:FBgn0000490"
                         /db_xref="PID:g157292"
                         /translation="MRAWLLLLAVLATFQTIVRVASTEDISQRFIAAIAPVAAHIPLA
                         SASGSGSGRSGSRSVGASTSTALAKAFNPFSEPASFSDSDKSHRSKTNKKPSKSDANR
                         .........
                         LGYDAYYCHGKCPFPLADHFNSTNHAVVQTLVNNMNPGKVPKACCVPTQLDSVAMLYL
                         NDQSTVVLKNYQEMTVVGCGCR"
    BASE COUNT     1170 a    1078 c     956 g    797 t
    ORIGIN
            1 gtcgttcaac agcgctgatc gagtttaaat ctataccgaa atgagcggcg gaaagtgagc
           61 cacttggcgt gaacccaaag ctttcgagga aaattctcgg acccccatat acaaatatcg
          121 gaaaaagtat cgaacagttt cgcgacgcga agcgttaaga tcgccaaaag atctccgtgc
          181 ggaaacaaag aaattgaggc actattaaga gattgttgtt gtgcgcgagt gtgtgtcttc
          241 agctgggtgt gtggaatgtc aactgacggg ttgtaaaggg aaaccctgaa atccgaacgg
          301 ccagccaaag caaataaagc tgtgaatacg aattaagtac aacaaacagt tactgaaaca
          361 gatacagatt cggattcgaa tagagaaaca gatactggag atgcccccag aaacaattca
          421 attgcaaata tagtgcgttg cgcgagtgcc agtggaaaaa tatgtggatt acctgcgaac
          481 cgtccgccca aggagccgcc gggtgacagg tgtatccccc aggataccaa cccgagccca
          541 gaccgagatc cacatccaga tcccgaccgc agggtgccag tgtgtcatgt gccgcggcat
          601 accgaccgca gccacatcta ccgaccaggt gcgcctcgaa tgcggcaaca caattttcaa
              ..........
         3841 aactgtataa acaaaacgta tgccctataa atatatgaat aactatctac atcgttatgc
         3901 gttctaagct aagctcgaat aaatccgtac acgttaatta atctagaatc gtaagaccta
         3961 acgcgtaagc tcagcatgtt ggataaaatta atagaaacga g
    //
```

Fig. 2.2. Examples of sequence database entries for (a) GenBank and (b) SWISS-PROT.

large-scale research projects. There are numerous genome databases, some of which are not open to the public but are kept within individual research groups. Genome databases are species specific, each containing different types of data such as genetic maps, physical maps, nucleotide sequences, and amino acid sequences for a given species. This is in clear contrast to the public database that is data type specific, con-

```
(b) ID    DECA_DROME      STANDARD;       PRT;     588 AA.
    AC    P07713;
    DT    01-APR-1988 (REL. 07, CREATED)
    DT    01-APR-1988 (REL. 07, LAST SEQUENCE UPDATE)
    DT    01-FEB-1995 (REL. 31, LAST ANNOTATION UPDATE)
    DE    DECAPENTAPLEGIC PROTEIN PRECURSOR (DPP-C PROTEIN).
    GN    DPP.
    OS    DROSOPHILA MELANOGASTER (FRUIT FLY).
    OC    EUKARYOTA; METAZOA; ARTHROPODA; INSECTA; DIPTERA.
    RN    [1]
    RP    SEQUENCE FROM N.A.
    RM    87090408
    RA    PADGETT R.W., ST JOHNSTON R.D., GELBART W.M.;
    RL    NATURE 325:81-84(1987).
    RN    [2]
    RP    CHARACTERIZATION, AND SEQUENCE OF 457-476.
    RM    90258853
    RA    PANGANIBAN G.E.F., RASHKA K.E., NEITZEL M.D., HOFFMANN F.M.;
    RL    MOL. CELL. BIOL. 10:2669-2677(1990).
    CC    -!- FUNCTION: DPP IS REQUIRED FOR THE PROPER DEVELOPMENT OF THE
    CC        EMBRYONIC DORSAL HYPODERM, FOR VIABILITY OF LARVAE AND FOR CELL
    CC        VIABILITY OF THE EPITHELIAL CELLS IN THE IMAGINAL DISKS.
    CC    -!- SUBUNIT: HOMODIMER, DISULFIDE-LINKED.
    CC    -!- SIMILARITY: TO OTHER GROWTH FACTORS OF THE TGF-BETA FAMILY.
    DR    EMBL; M30116; DMDPPC.
    DR    PIR; A26158; A26158.
    DR    HSSP; P08112; 1TFG.
    DR    FLYBASE; FBGN0000490; DPP.
    DR    PROSITE; PS00250; TGF_BETA.
    KW    GROWTH FACTOR; DIFFERENTIATION; SIGNAL.
    FT    SIGNAL        1       ?       POTENTIAL.
    FT    PROPEP        ?     456
    FT    CHAIN       457     588       DECAPENTAPLEGIC PROTEIN.
    FT    DISULFID    487     553       BY SIMILARITY.
    FT    DISULFID    516     585       BY SIMILARITY.
    FT    DISULFID    520     587       BY SIMILARITY.
    FT    DISULFID    552     552       INTERCHAIN (BY SIMILARITY).
    FT    CARBOHYD    120     120       POTENTIAL.
    FT    CARBOHYD    342     342       POTENTIAL.
    FT    CARBOHYD    377     377       POTENTIAL.
    FT    CARBOHYD    529     529       POTENTIAL.
    SQ    SEQUENCE    588 AA;   65850 MW;   1768420 CN;
          MRAWLLLLAV LATFQTIVRV ASTEDISQRF IAAIAPVAAH IPLASASGSG SGRSGSRSVG
          ASTSTALAKA FNPFSEPASF SDSDKSHRSK TNKKPSKSDA NRQFNEVHKP RTDQLENSKN
          KSKQLVNKPN HNKMAVKEQR SHHKKSHHHR SHQPKQASAS TESHQSSSIE SIFVEEPTLV
          LDREVASINV PANAKAIIAE QGPSTYSKEA LIKDKLKPDP STLVEIEKSL LSLFNMKRPP
          KIDRSKIIIP EPMKKLYAEI MGHELDSVNI PKPGLLTKSA NTVRSFTHKD SKIDDRFPHH
          HRFRLHFDVK SIPADEKLKA AELQLTRDAL SQQVVASRSS ANRTRYQVLV YDITRVGVRG
          QREPSYLLLD TKTVRLNSTD TVSLDVQPAV DRWLASPQRN YGLLVEVRTV RSLKPAPHHH
          VRLRRSADEA HERWQHKQPL LFTYTDDGRH KARSIRDVSG GEGGGKGGRN KRHARRPTRR
          KNHDDTCRRH SLYVDFSDVG WDDWIVAPLG YDAYYCHGKC PFPLADHFNS TNHAVVQTLV
          NNMNPGKVPK ACCVPTQLDS VAMLYLNDQS TVVLKNYQEM TVVGCGCR
    //
```

taining, for example, just nucleotide sequences for all species. The data organization in genome databases is biological, rather than bibliographic; in essence, the genome database represents genome structures at different resolutions and genome functions at different levels.

The highest resolution map of the genome structure is the nucleotide sequence of the complete genome, which can be determined by high-throughput DNA sequencing methods. Genes coding for proteins and RNAs can then be identified by gene finding methods and their functions can partly be deduced by sequence similarity search methods (see *Gene finding and functional predictions* in Section 3.2).

In many genome databases genes are hierarchically classified according to their functions. An attempt to establish such a functional hierarchy was first made for *Escherichia coli* by Monica Riley. Table 2.4 is her initial classification of *E. coli* gene functions showing only the top two levels of the hierarchy. The classification is inevitably biased toward what is already known, especially metabolism since that is the best characterized part of cellular functions.

The genome sequencing project will uncover, at least, a complete catalogue of genes for a given organism. However, the functions of roughly half of the genes in every genome that has been sequenced remain unknown because either no similarities are found in the databases or similarities are found only to unknown sequences. Even though the entire sequence information is deposited and made available in the public sequence databases, many genome databases still remain

Table 2.4. Functional classification of *E. coli* genes according to Monica Riley

I.   Intermediary metabolism
    A. Degradation
    B. Central intermediary metabolism
    C. Respiration (aerobic and anaerobic)
    D. Fermentation
    E. ATP-proton motive force interconversion
    F. Broad regulatory functions
II.   Biosynthesis of small molecules
    A. Amino acids
    B. Nucleotides
    C. Sugars and sugar nucleotides
    D. Cofactors, prosthetic groups, electron carriers
    E. Fatty acids and lipids
    F. Polyamines
III. Macromolecule metabolism
    A. Synthesis and modification
    B. Degradation of macromolecules
IV   Cell structure
    A. Membrane components
    B. Murein sacculus
    C. Surface polysaccharides and antigens
    D. Surface structures
V   Cellular processes
    A. Transport/binding proteins
    B. Cell division
    C. Chemotaxis and mobility
    D. Protein secretion
    E. Osmotic adaptation
VI.   Other functions
    A. Cryptic genes
    B. Phage-related functions and prophages
    C. Colicin-related functions
    D. Plasmid-related functions
    E. Drug/analog sensitivity
    F. Radiation sensitivity
    G. DNA sites
    H. Adaptations to atypical conditions

useful resources because they keep adding and revising functional information of individual genes. In selected genomes, coordinated efforts are undertaken to investigate functional genomics, such as by observing gene expression profiles at the mRNA level (transcriptome) and at the protein level (proteome). The genome databases will continue to evolve by incorporating such new types of expression data generated by the emerging experimental technologies of DNA chips and protein chips.

2.2   Informatics technologies

Relational databases

A database is a computerized collection of data with additional notions. It serves a practical purpose of information retrieval, it is usually shared by many users, and it is organized with a predefined set of data items. It is also assumed to be managed by a computer program called the database management system. Hence, the data items must be defined with consideration of computer processing—what types of records constitute a database, what kinds of attributes exist in a record, what values are possible for an attribute, and so on. The schema is the specification of a logical structure of the database based on such data definition. The format of the database, as in the flat file format, can also be considered the schema in a broad sense.

When designing a database, it is necessary to somehow model the real world yet make the data accessible in a computer. The data model is a conceptual and operational framework for this modelling. Traditional data models include: the hierarchical data model, the network data model, and the relational data model. The relational data model is intuitive and easily comprehensible. All the data are organized in tables; namely, the world is viewed as a collection of tables. The relational data model was first proposed by E. F. Codd in 1970. The relational database management system (RDBMS) has since become most popular in commercially available systems.

In the relational database a table is also called a relation. Each row of the table corresponds to a record, also known as a tuple, and each column corresponds to an attribute of the record. Strictly speaking, a relation is a set; there is no notion of the position of a record within a table or the position of a column within a record. Figure 2.3 shows an example of a simple bibliographic database. Here a record in one table contains the citation information that consists of the reference identifier (MUID), journal name, volume, pages, and published year. Suppose that the user wishes to search references that were published in a given year. The search would create a smaller table by extracting matching records from the original table. The process of extracting given rows is called 'selection' in the relational database. In contrast, the process of extracting given columns is called 'projection'. In Fig. 2.3 there is another table containing the author information. In order to find references that were published in a given journal by a given author, it is necessary to search both tables at a time. This is made possible by the 'join' operation where the records that contain matching values for the specified attribute will be merged into one table. In this example, the matching values for the attribute MUID are used to combine records from the two tables into one longer record in the resulting table.

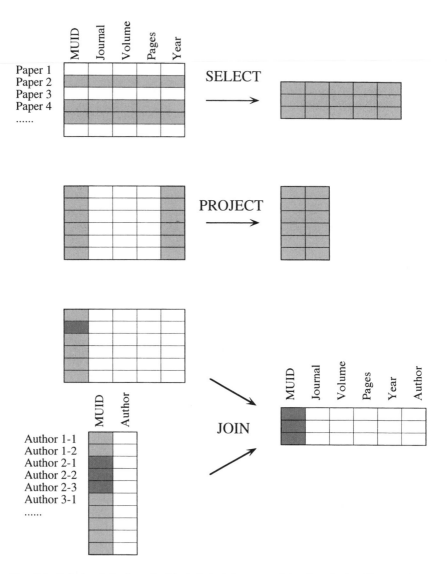

Fig. 2.3. Relational database. A table (relation) is a set and the three basic table operations shown here are extensions of the standard set operations.

The information retrieval in the relational database is thus realized by table operations, which are extensions of set operations as formulated by Codd. The mathematical basis of the relational database is called relational algebra, which provides a sound foundation for developing practical database management systems. On the other hand, the relational database is often criticized for its lack of flexibility to represent real world data in natural ways. In order to conform to the normalized form of the table, it is often required to split the data into many tables, which necessitates the increase of costly join operations. In the example of Fig. 2.3, MUID is a unique primary key in the citation table, but in the joined table multiple records appear for the same MUID whenever the author table contains multiple authors for the same MUID. Intuitively, the two tables would be better organized in a single hierarchical table where multiple author records appear under the same citation record. The flat-file format of the sequence databases allows such a hierarchy; multiple REFERENCE records appear under the same LOCUS entry in GenBank. This and other types of data structures are far better represented by the object-oriented database (see *Object-oriented databases* below).

In general the database queries are made by the database language. In the relational database the standard database language is SQL (Structured Query Language). A basic form of SQL is:

> SELECT   list-of-attributes
>
> FROM   list-of-relations
>
> WHERE   condition

For example, the selection operation in Fig. 2.3 may be written as:

SELECT * FROM CITATION_TABLE WHERE PUBLISHD_YEAR='1995'

where * represents all attributes in the table. When the list of attributes is specified, the projection operation will take place. The join operation can be invoked by simply specifying multiple tables in the list-of-relations and associated WHERE conditions. SQL is a declarative language; it is like a natural language expressing what the user wishes to obtain rather than how it should be obtained. This is due to the underlying mathematical foundation of relational algebra that the relational database utilizes. In other types of databases, such as those based on the network data model, the query language tends to be procedural; it is more like a programming language expressing the procedure for the specific data structures involved.

Deductive databases

Linkage analysis is a powerful tool that identifies disease genes by analysing pedigrees (family trees) for any correlations between affected individuals and specific markers in the human genome (see Fig. 1.11). A family tree shown in Fig. 2.4 is a complex data type that is difficult to represent by the relational database. In contrast, this type of problem can better be addressed by using the deductive database. The basic data item in the family tree is a parent–child relation, and by recursively using parent–child relations it is possible to identify individuals in the family tree. Let us first briefly overview the underlying concepts in predicate logic.

A predicate is a logical expression whose value may be true or false depending on the values of the specific variables that it contains. This is in contrast with a proposition, which is like a logical constant where the value of either true or false is predetermined. For example, given two variables X and Y, the statement 'X is a parent of Y' is a predicate. A predicate is a type of logical expression that can be combined with logical operators to form more complex logical expressions. The

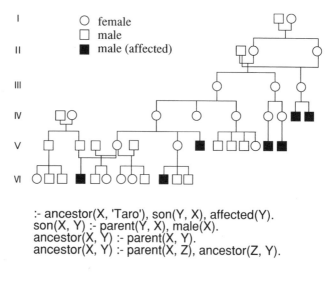

:- ancestor(X, 'Taro'), son(Y, X), affected(Y).
son(X, Y) :- parent(Y, X), male(X).
ancestor(X, Y) :- parent(X, Y).
ancestor(X, Y) :- parent(X, Z), ancestor(Z, Y).

Individual	Sex	Father	Mother	Affected
Taro	male	Ichitaro	Hanako	yes
Jiro	male	Ichitaro	Hanako	no
....				

Fig. 2.4. Deductive database. The data in the family tree is represented and manipulated in a deductive database, which consists of a relational database and a logic programming interface.

logical operators include: disjunction (or), conjunction (and), negation (not), implication (if…then), universal operator (for all), and existential operator (there exists). There are variations of predicate logic with different sets of logical expressions, but the first-order predicate logic is most relevant in computer science. The three types of logical expressions in the first-order predicate logic:

$$P \leftarrow Q_1, Q_2, \ldots, Q_n$$
$$\leftarrow Q_1, Q_2, \ldots, Q_n$$
$$P \leftarrow$$

are called Horn clauses. The first type is a rule clause, representing: 'if Q_1 and Q_2 and … and Q_n then P'. The third type is a unitary clause; unitary clauses without any variables are called facts. The second type is a goal clause that actually represents a query. The logic programming language Prolog was developed by R. A. Kowalski based on these Horn clauses.

In the example of Fig. 2.4, the unitary clauses 'parent (X, Y)' and 'ancestor (X, Y)' represent, respectively, 'X is a parent of Y' and 'X is an ancestor of Y'. Because a parent is included in an ancestor, the rule:

$$\text{ancestor } (X, Y) \leftarrow \text{parent } (X, Y)$$

can be defined. Furthermore, because a parent of the ancestor also belongs to the ancestor, a recursive rule:

$$\text{ancestor } (X, Y) \leftarrow \text{parent } (X, Z), \text{ancestor } (Z, Y)$$

can also be established. The question of asking Taro's parents can be written in the second form of the Horn clause:

$$\leftarrow \text{parent } (X, \text{'Taro'})$$

The facts of parent–child relations could be written in the third form of the Horn clause, but in actual implementations they are usually stored in the relational database. Figure 2.4 illustrates an implementation of a deductive database that consists of a relational database for storing facts and logic programming for description of rules and queries.

Using the predicate 'person (Individual, Sex, Father, Mother, Affected)' the information in the table of Fig. 2.4 can be written in the third type of Horn clause:

person ('Taro', 'male', 'Ichitaro', 'Hanako', 'yes')

person ('Jiro', 'male', 'Ichitaro', 'Hanako', 'no')

…

This is the definition of the relational database in terms of relational calculus rather than relational algebra. Thus, the deductive database is a natural extension of the relational database by incorporating the other two types of Horn clauses in the first-order predicate logic. Originally, Prolog and other logic programming languages were used to solve search and inference problems in artificial intelligence. It is interesting to note that because of the deductive database the concept of logic programming was naturally connected to the database problem.

Object-oriented databases

Another concept in the programming language was also found to be extremely useful in the database problems. It is the concept of object-oriented programming, which was first implemented in Smalltalk by A. Kay. Although the deductive database is an improvement over the relational database in its ability to better represent data, it is in no way sufficient to cope with the diversity and complexity of molecular biology data. For a more flexible representation of real world data, the object-oriented database is generally considered to be a better choice over the relational database. In general, however, a logic-based system has more flexibility for computation of data. A query in an object-oriented database tends to be procedural, while a query in a relational/deductive database can be done in a declarative way as in SQL. This is certainly an advantage for the relational/deductive database, though the trade-off is a formal, inflexible representation of data.

An object is a self-contained module consisting of a data item and an associated procedure for processing. An object is encapsulated; namely, it is an abstract data type whose internal form is hidden behind a set of routines, called methods, that operate on the data item. By message passing, methods are invoked to process an object. Each object is distinguished by an identifier; objects with identical characteristics are considered different if identifiers are different. This object identity can be compared with the relational database where a record is distinguished by the values. Remember that there is no notion of position of a record within a table. Two records with identical values for all the attributes cannot exist in the same table. Therefore, the relational database and the deductive database can be termed value-oriented. Objects with common properties and procedures are grouped into a class; thus, objects are instances of a particular class. Classes are organized in a hierarchy with the properties and procedures defined in an upper class being automatically inherited down the hierarchy to lower classes. In the object-oriented programming, objects are independent program modules that serve as the building blocks of many applications.

In the object-oriented database the data are stored as objects and managed by an object-oriented database management system. Because of the abstract data type the object-oriented database has far greater flexibility for data representation than the relational database or the deductive database. However, it has less computational capacity because it lacks any mathematical or logical foundation such as relational algebra or predicate logic. In practical terms this means that the relational database management system is generally more robust against errors and inconsistencies, thus assuring data integrity. Consequently, many of the existing molecular biology databases are still organized as relational databases with additions of object-oriented programs and user interfaces.

Figure 2.5 illustrates the concept of an object-oriented database for genes. A gene may represent a number of different data items depending on the domain of interests. For example, a gene can be a chromosomal location, a nucleotide sequence, a translated amino acid sequence, an encoded 3D protein structure, a role in the biochemical pathway, a time- and space-dependent expression profile, or a mutation pattern that is related to a disease. In order to understand a given gene or a given set of genes it is probably necessary to compare the gene(s) with many genes in different organisms. A search could be conducted to look for similarity of map positions called synteny, sequence similarities both at the nucleotide and amino acid sequence levels, 3D structural similarity, pathway similarity, and similarities of gene expression profiles and genomic mutation patterns. These similarity searches could be performed using different computational procedures for different data types, and are thus well-suited to the concepts of encapsulation and class inheritance in the object-oriented database.

The development of the aforementioned database technologies is summarized in Fig. 2.6. It is fair to say that though the deductive database is not as popular as the object-oriented database, its capability to logically deduce new data from stored data is particularly useful in molecular biology. This allows for a systematic survey of new knowledge to be generated from stored knowledge, which is somewhat similar to a human reasoning step. Attempts to combine the object-oriented and deductive databases have been made. In practice, however, the relational database will continue to be the basic database management system, with the flavours of object-oriented programming and logic programming added to greatly enhance the capacity of the relational database.

Link-based integration

Perhaps, the most striking aspect of molecular biology databases is the rate of increase both in terms of the size of the sequence databases and the number of other divergent databases. In order to integrate different types of data into a single relational database, it is necessary to develop a unified schema. This is not a trivial task because different data are provided by different sources and different sources may occasionally change their own schemata. Furthermore, except for trivial facts such as bibliographic information or nucleotide sequence information, what is represented in the record and its attributes contains semantics that may not be easily reconciled among different databases. For example, the attribute 'function' for the record 'gene' could contain various pieces of information depending on what is meant by 'gene' and whether the level of interest is molecular or phenotypic.

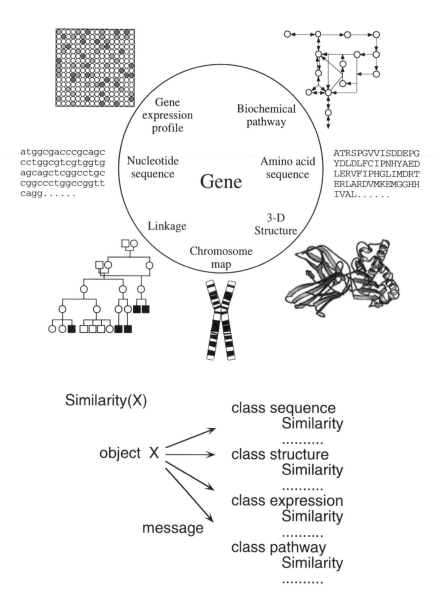

Fig. 2.5. Object-oriented database. The concept of similarity is implemented in an object-oriented database which incorporates many different aspects of genes.

In contrast to the strong integration of a unified schema in the relational database, a weak integration called link-based integration is far more practical. What is agreed upon here is minimal—only that the database is a collection of entries, as

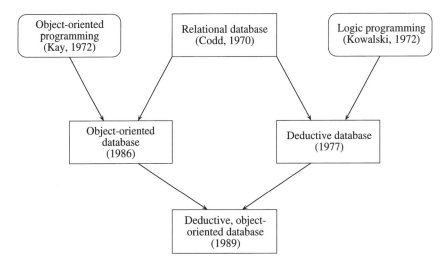

Fig. 2.6. A history of database technology development.

in the flat-file view of molecular biology databases. Without bothering with the fine structure of each entry it is possible to specify whether two entries in different databases are, in any way, related or not. It is customary, or even required, to incorporate such cross-reference information into the current molecular biology databases.

In the flat-file view, any entry in the web of molecular biology databases can be specified by the combination of the database name and the unique identifier of an entry (the entry name or the primary accession number):

database:entry

The link information between two entries is a binary relation in the form of:

database1:entry1 → database2:entry2

which represents, for example, the relation between the reference information and the reported nucleotide sequence information or between the nucleotide sequence information and the translated amino acid sequence information.

We have already seen an example of the binary relation, namely, the parent–child relation in the deductive database. By repeatedly using multiple parent–child relations a group of ancestors can be identified. Similarly, by repeatedly using relations between two database entries, including their reverse relations, a number of additional links can be computationally derived. This computation is useful because the cross-reference efforts in major biological databases are not comprehensive; some databases such as SWISS-PROT provide ample links while others do not. Figure 2.7 shows the DBGET/LinkDB system under the Japanese GenomeNet service, a practical system of link–based integration.

Knowledge base

As was mentioned first in this chapter, knowledge is different from data in that new knowledge can be generated from existing knowledge. In our definition, a knowledge base is different from a database in that it contains the intrinsic capacity to compute new knowledge from stored knowledge. Though the relational database might simply be a static collection of facts that can only be used for information retrieval, the deductive database is a dynamic collection of facts and rules that can also be used for logical inference. Thus, the deductive database is a typical knowledge base. Another type of knowledge base is an expert system, or more specifically a production system, which is a collection of 'if–then' rules with an inference

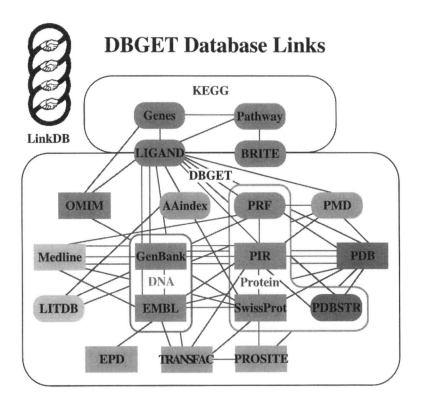

Fig. 2.7. A link-based integration of molecular biology databases in the DBGET/LinkDB system at GenomeNet (http://www.genome.ad.jp/). The lines indicate that the cross-references are given by the original databases.

engine (see *Expert system for protein sorting prediction* in Section 3.2).

Given two binary relations: 'A → B' and 'B → C', a third binary relation 'A → C' can be deduced. This syllogism is the basis of sequence similarity search. Sequence A is similar to sequence B. Sequence B is known to have biological function C. Therefore, sequence A is also likely to have biological function C. While all the known sequence data are stored in the sequence databases and the similarity links A → B can be computed by sequence comparison algorithms, the lack of appropriate functional databases makes the entire process difficult to computerize. Though it may be possible to describe function C for humans to understand, that is not sufficient for C to be used in further computerized reasoning steps.

In the KEGG (Kyoto encyclopedia of genes and genomes) system (Table 2.3) that is also available under the Japanese GenomeNet service, an organism is considered a database of genes. Any gene or gene product in an organism is specified by:

organism:gene

and any relation between genes or gene products is represented by:

organism1:gene1 → organism2:gene2

This binary relation would represent, for example, orthologous genes between two organisms or a protein–protein interaction in the signal transduction pathway of an organism. Links are also made to other databases:

organism:gene → database:entry

An example is the assignment of an EC number to a gene product, which is a sequence–function relationship with a standardized definition of function.

A piece of nucleotide sequence data is not knowledge *per se* unless it is linked to another data item. Generally speaking, biological entities and relationships among them together constitute knowledge in biology. An ontology is a formal specification of how to represent entities and their relationships based on an abstract conceptualization of the world. In the ontology of traditional molecular biology databases, the emphasis has been given to genes and molecules, which certainly are the basic building blocks of life. However, it is emphasized here that the interactions and relationships among them are not mere attributes of these building blocks, but actually form the basic objects and concepts, which are as important as the genes and molecules themselves. By conforming to the simplest way of representing relationships, namely, the binary relation, it is expected that logical inference, such as the interpretation of sequence data, would better be computerized.

World Wide Web

The concept of hyperlinks in the World Wide Web (WWW) has very much in common with the link-based integration of molecular biology databases. In practice too, the proliferation of the WWW in the mid 1990s dramatically enhanced the usability of molecular biology databases. The address in the WWW is a combination of the machine name and the file name

machine/file

such as 'www.genome.ad.jp/kegg/kegg1.html' where 'www.genome.ad.jp' is a machine name and 'kegg/kegg1.html' is a file name. The uniform access locator (URL) of the WWW then takes the form of 'http://www.genome.ad.jp/kegg/kegg1.html' where 'http' is an access method. The hyperlink is a binary relation represented by

machine1/file1 → machine2/file2

This is again based on a flat-file view; first, a machine is a collection of files and second, related files in different machines can be linked without going into the detail of how each file (page) is organized. This is also object-oriented; an encapsulated object that is both data and procedure can be accessed by a pre-defined method.

The WWW is based on a server–client mechanism. Once a network browser program is installed on the client side, i.e. in the user's machine, the user can access a virtually limitless number of server machines over the Internet and make use of enormous amounts of database and computational resources. Browser programs generally handle not only text files but also GIF image and other standard multimedia files. They can also be enhanced to process non-standard multimedia files by installing plug-in software. Furthermore, they can interpret and run programs such as Java scripts and Java applets that are dynamically transferred from the server machine. While the majority of molecular biology data is currently represented by text, such multimedia capability of the WWW has the potential to transform the database's organization. As summarized in Table 2.5 the GenomeNet databases are provided in multimedia, including the protein structure by 3D graphics, the chemical structure by 2D graphics, and the biochemical pathway map by GIF image.

Computer graphics

Graphical representation is indispensable for understanding specific types of data in molecular biology, especially the three-dimensional structural data for biological macromolecules of proteins, RNAs, and DNAs. The 3D structure is represented by the numerical values of atomic coordinates in the databases or in the results of computations. It is then visualized and manipulated by computer graphics called

Table 2.5. Multimedia in GenomeNet

Data type	Database	Media
Nucleic acid sequences	GenBank, EMBL	Text
Protein sequences	SWISS-PROT, PIR, PRF	Text
3D molecular structures	PDB	Text, 3D graphics
Sequence motifs	EPD, TRANSFAC, PROSITE	Text, 3D graphics
Chemical reactions	LIGAND/ENZYME	Text
Chemical compounds	LIGAND/COMPOUND	Text, image, 2D graphics
Biochemical pathways	KEGG/PATHWAY	Image, Java applet
Gene catalogues	KEGG/GENES	Text
Genomes	KEGG/GENOME	Text, image, Java applet
Expression profiles	KEGG/EXPRESSION	Image, Java applet
Genetic diseases	OMIM	Text
Amino acid mutations	PMD	Text
Amino acid indices	AAindex	Text
Literature	Medline, LITDB	Text
Database links	LinkDB	Text

3D molecular graphics. The existing 3D molecular graphics software is capable of handling 3D structural databases and is often integrated with computational tools in quantum chemistry, molecular mechanics, and molecular dynamics. It may also have additional capabilities to do molecular modelling, such as to perform site-directed mutagenesis *in silico* or to examine the docking of a protein and a ligand.

Figure 2.8 shows some typical graphic representations of how the 3D structures of biological macromolecules are visualized. In the wire-frame model, which is also called the skeleton model, the atoms and covalent bonds are represented by dots and lines, respectively. The ball-and-stick model is similar but instead of dots, spheres of finite sizes are used for the atoms. Both models are suited to visualize the network of atomic connections. They can be simplified to contain only the carbon and nitrogen atoms in the polypeptide backbone or even only the α-carbon atoms, so that the overall protein folding pattern can be better captured. A more aesthetic representation is the ribbon model, which emphasizes the arrangement of the secondary structures of α-helices and β-strands. When the atoms are represented by respective van der Waals radii, the space-filling model, also called the CPK model, is generated. It presents a feel for the overall shape and volume of the molecule. In contrast to a sense of physical touch, the dot surface model utilizes a collection of dots to visualize other surface properties like the electrostatic potential or the accessible surface area for water molecules. All of these representations can be in stereo, where two tilted images are generated and viewed with trained eyes or with a special glass.

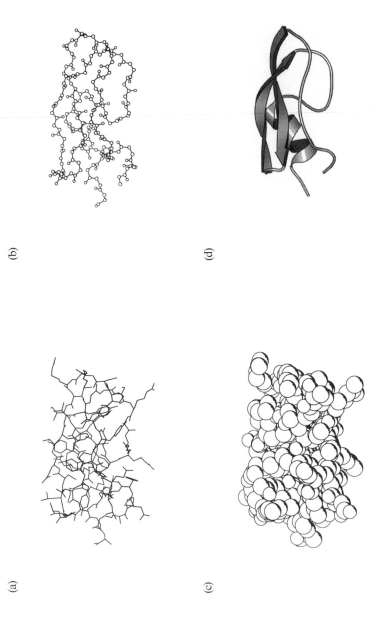

(b)

(d)

(a)

(c)

Fig. 2.8. Representations of the three–dimensional structure of pancreatic trypsin inhibitor (PDB:4PTI) by computer graphics: (a) wire–frame model, (b) ball-and-stick model (main chain only), (c) space-filling model, and (d) ribbon model.

2.3   New generation of molecular biology databases

Elements and compounds

By definition living organisms consist of organic compounds that are derived mostly from hydrogen (H), carbon (C), nitrogen (N), and oxygen (O). Together with phosphorus (P) in the nitrogen group and sulfur (S) in the oxygen group in the periodic table (Fig. 2.9) they comprise the six essential elements in life. The two alkali metals, sodium (Na) and potassium (K), the two alkaline earth metals, magnesium (Mg) and calcium (Ca), and the halogen chlorine (Cl) are also biologically important. There are additional trace elements, including the transition metals, manganese (Mn), iron (Fe), cobalt (Co), nickel (Ni), copper (Cu), zinc (Zn), and molybdenum (Mo), that are often actively transported into the cell and incorporated into biomolecules.

   Life cannot be sustained without sources of carbon and energy. The distinction between autotrophs and heterotrophs lies in the ability of the autotrophs to depend only on carbon dioxide (CO_2) for the carbon source and either light or oxidizable inorganic compounds for the energy source. In contrast, heterotrophs require organic compounds as the sources of both carbon and energy, and hence ultimately depend on autotrophs. Autotrophs perform the process of carbon dioxide (CO_2) fixation, during which CO_2 is converted to useful organic compounds. The energy is input by light in plants and photosynthesizing bacteria, or it is provided by oxidation of sulfur (S), sulfite (SO_3^{2-}), thiosulfate ($S_2O_3^{2-}$), hydrogen sulfide (H_2S), nitrite (NO_2^-), ammonia (NH_3), ferrous (Fe^{2+}) compounds, and other inorganic compounds in chemolithotrophic bacteria. Figure 2.10 shows important classes of

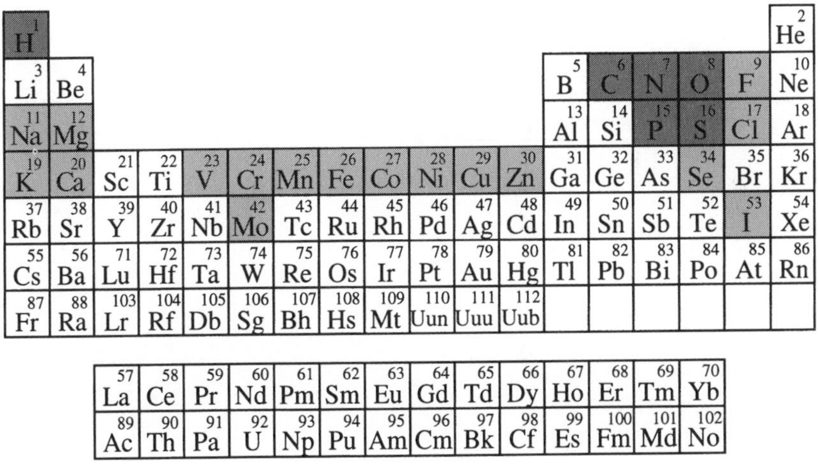

Fig. 2.9. The periodic table of chemical elements where the shaded elements are those normally found in biology.

H (hydrogen)	C (carbon)	N (nitrogen)	O (oxygen)
	CO_2 (carbon dioxide)	NO_3^- (nitrate)	
	HCO_3^- (hydrogen carbonate)	NO_2^- (nitrite)	

CH_4 (methane)

CH_3 (methyl group) COOH (carboxyl group) NH_3 (ammonia) H_2O (water)

R (alkyl group) R-COOH (carboxylic acid) NH_2 (amino group) OH (hydroxyl group)

NH_2-CHR-COOH (amino acid) R-NH_2 (amine) R-OH (alcohol) R-CHO (aldehyde)

R-O-R' (ether) R-CO-R' (ketone)

P (phosphorus)
PO_4^{3-} (phosphate)

S (sulfur)
SO_4^{2-} (sulfate)
SO_3^{2-} (sulfite)
$S_2O_3^{2-}$ (thiosulfate)

H_2S (hydrogen sulfide)

SH (sulfhydryl group)

R-SH (thiol)

R-S-S-R' (disulfide)

R-COO-R   (carboxylic acid ester such as fats)

HPO_3-O-R'   (phosphoric acid monoester such as phospholipids)

R-O-PO_2-O-R'   (phosphodiester bond in nucleic acids)

R-NH-CO-R'   (peptide bond in proteins)

R-S-CO-R'   (thioester such as acetyl-CoA)

Fig. 2.10. Biologically important classes of organic compounds derived from the six basic elements.

organic compounds in biology that are derived from different forms of inorganic compounds of the six essential elements. Phosphorus exists in biology as derivatives of phosphoric acid (H_3PO_4), especially as the source of high-energy bonds.

While knowledge of organic chemistry and its biological relevance is extremely important, the existing chemical databases, such as CAS registry, are proprietary and not suited for integration in the web of molecular biology databases. The LIGAND database (Table 2.3), organized as part of the KEGG system, attempts to fill in the gap. LIGAND contains information on chemical compounds and chemical reactions in living cells, as well as information on enzymes, with rich cross-references to the major molecular biology databases. LIGAND also provides a classification of biologically important compounds.

Amino acid indices

The 20 amino acids (Fig. 2.11) specified by the genetic code have different side chain constituents, which in turn give each amino acid different physical properties, such as shape and volume, and chemical reactivities. For example, the amino acids can be classified into hydrophobic or hydrophilic, charged or uncharged, with or without hydroxyl (OH) group, and so on. Furthermore, similar amino acids in one grouping may belong to different groups when categorized differently. The repertoire of these versatile building blocks makes it possible to generate the variety and specificity of protein three-dimensional structures and their biological functions. A large body of experimental and theoretical work has been performed to characterize physicochemical and biochemical properties of individual amino acids. The derived property is often represented by a set of 20 numerical values, which is called the amino acid index. By using a hydrophobicity index the amino acid sequence is converted to a numerical profile, as first proposed by Kyte and Doolittle in their hydropathy plot, from which a possible transmembrane segment can be identified. The Chou–Fasman method, which is the very first method for protein secondary structure prediction, is based on the analysis of statistical propensities of individual amino acids to form α-helices, β-sheets, and turns in known protein structures (see *Prediction of protein secondary structures* and *Prediction of transmembrane segments* in Section 3.2).

In addition to the properties of individual amino acids, the similarity relations between amino acids can be represented by a set of 20×20 numerical values, which is called the amino acid mutation matrix or the amino acid similarity score matrix. The matrix is usually symmetric so that it consists of 210 numerical values. The major use of the matrix is as a measure for optimization in protein sequence alignments and similarity searches. Dayhoff and coworkers were the first to compile such a mutation matrix. They constructed phylogenetic trees from groups of closely related proteins and collected the data of accepted point mutations (PAMs) per 100 residues. Their log-odds matrix called the PAM matrix (Table 2.6) is still widely used as a scoring scheme (see *Global alignment* in Section 3.1).

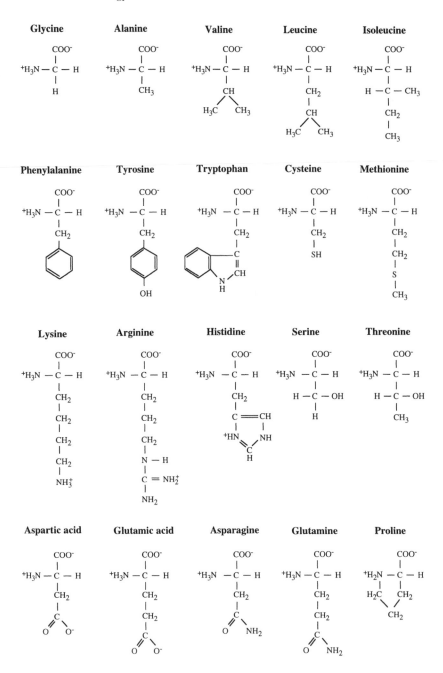

Fig. 2.11. The twenty common amino acids.

Table 2.6. The PAM-250 mutation matrix

	Ala	Arg	Asn	Asp	Cys	Gln	Glu	Gly	His	Ile	Leu	Lys	Met	Phe	Pro	Ser	Thr	Trp	Tyr	Val
Ala	2																			
Arg	-2	6																		
Asn	0	0	2																	
Asp	0	-1	2	4																
Cys	-2	-4	-4	-5	12															
Gln	0	1	1	2	-5	4														
Glu	0	-1	1	3	-5	2	4													
Gly	1	-3	0	1	-3	-1	0	5												
His	-1	2	2	1	-3	3	1	-2	6											
Ile	-1	-2	-2	-2	-2	-2	-2	-3	-2	5										
Leu	-2	-3	-3	-4	-6	-2	-3	-4	-2	2	6									
Lys	-1	3	1	0	-5	1	0	-2	0	-2	-3	5								
Met	-1	0	-2	-3	-5	-1	-2	-3	-2	2	4	0	6							
Phe	-4	-4	-4	-6	-4	-5	-5	-5	-2	1	2	-5	0	9						
Pro	1	0	-1	-1	-3	0	-1	-1	0	-2	-3	-1	-2	-5	6					
Ser	1	0	1	0	0	-1	0	1	-1	-1	-3	0	-2	-3	1	2				
Thr	1	-1	0	0	-2	-1	0	0	-1	0	-2	0	-1	-3	0	1	3			
Trp	-6	2	-4	-7	-8	-5	-7	-7	-3	-5	-2	-3	-4	0	-6	-2	-5	17		
Tyr	-3	-4	-2	-4	0	-4	-4	-5	0	-1	-1	-4	-2	7	-5	-3	-3	0	10	
Val	0	-2	-2	-2	-2	-2	-2	-1	-2	4	2	-2	2	-1	-1	-1	0	-6	-2	4
	Ala	Arg	Asn	Asp	Cys	Gln	Glu	Gly	His	Ile	Leu	Lys	Met	Phe	Pro	Ser	Thr	Trp	Tyr	Val

The AAindex database (Table 2.3) is a collection and a classification of published amino acid indices and mutation matrices. According to this database there are five major classes of amino acid indices: α-helix and turn propensities, β-strand propensity, hydrophobicity, amino acid composition, and physicochemical properties such as bulkiness. It is interesting to note that the α-propensity and the turn propensity are well related, possibly suggesting common structural features and reflecting the tendency of helices to appear on the surface of a globular protein. The β-propensity is linked closely to the hydrophobicity, which is consistent with the tendency of β-sheets to be buried inside the globule. The grouping of published amino acid mutation matrices is not clear-cut, but those obtained from families of similar sequences and those obtained from structural analysis tend to be different. In general, an amino acid mutation matrix contains multiple properties of amino acids—for example, the PAM matrix largely reflects the volume and hydrophobicity of individual amino acids.

Protein families and sequence motifs

The collection of amino acid sequence data in the form of primary sequence databases (Tables 2.1 and 2.2) contains rich information of protein structure, function, and evolution. Because a group of proteins with similar amino acid sequences often adopts similar 3D structures, shares common biological functions, and reflects evolutionary relationships, there have been attempts to organize secondary databases of protein families. An original attempt was made in the late 1960s by Dayhoff and coworkers who established the concept of superfamilies, which is still part of the activities of the PIR database. According to their definition, a subfamily is a group of almost identical sequences where a pair of aligned amino acid sequences contains 90% or more identical amino acids. A family is a group of proteins with 50% or more identical amino acids in each pair, whose functions are generally very similar. A superfamily is a broad classification where each pair of proteins may contain any trace of significant sequence similarity that is measured not by the amino acid identity but by the PAM mutation matrix. Proteins in the same superfamily may be evolutionarily related but may have somewhat different functions.

Originally the purpose of sequence comparison was to analyse molecular evolution, but it is now primarily used for understanding functional implications, i.e. extending sequence similarity to functional similarity. Even when the overall sequence similarity is marginal, or in the so-called twilight zone, two sequences may sometimes share localized regions of conserved amino acids called sequence motifs. These regions are functionally important sites of protein molecules, and are often sites for interactions with other molecules. As shown in Fig. 2.12 sequence motifs can be represented in different ways. The consensus sequence pattern is most easily comprehensible but least effective in representing the complexity of biology. The pattern basically specifies the order of conserved amino acids by allowing alternatives and gaps in a form similar to the UNIX regular expression (see *Formal grammar* in Section 3.2). A number of consensus sequence patterns have been identified by multiple sequence alignment and visual inspection of individual protein families. The patterns reported in the literature along with their functional implications are computerized in motif libraries such as PROSITE (Table 2.3). Since the

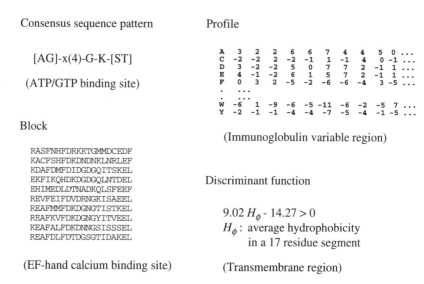

Consensus sequence pattern

[AG]-x(4)-G-K-[ST]

(ATP/GTP binding site)

Block

```
RASFNHFDRKKTGMMDCEDF
KACFSHFDKDNDNKLNRLEF
KDAFDMFDIDGDGQITSKEL
EKFIKQHDKDGDGQLNTDEL
EHIMEDLDTNADKQLSFEEF
REVFEIFDVDRNGKISAEEL
REAFMMFDKDGNGTISTKEL
REAFKVFDKDGNGYITVEEL
KEAFALFDKDNNGSISSSEL
REAFDLFDTDGSGTIDAKEL
```

(EF-hand calcium binding site)

Profile

```
A    3   2   2   6   6   7   4   4   5   0 ...
C   -2  -2   2  -2  -1   1  -1   4   0  -1 ...
D    3  -2  -2   5   0   7   7   2  -1   1 ...
E    4  -1  -2   6   1   5   7   2  -1   1 ...
F    0   3   2  -5  -2  -6  -6  -4   3  -5 ...
.    ...
.    ...
W   -6   1  -9  -6  -5 -11  -6  -2  -5   7 ...
Y   -2  -1  -1  -4  -4  -7  -5  -4  -1  -5 ...
```

(Immunoglobulin variable region)

Discriminant function

$9.02 H_\phi - 14.27 > 0$

H_ϕ: average hydrophobicity
in a 17 residue segment

(Transmembrane region)

Fig. 2.12. Different representations of sequence motifs: the consensus sequence pattern, the block, the profile, and the discriminant function.

motif library is a dictionary of sequence–function relationships, a search performed against the library is useful for functional interpretation of newly determined sequences, especially when no global sequence similarity can be found by sequence database searches (see *Problems in sequence analysis* in Section 3.1).

Although the consensus sequence pattern representation allows alternative amino acids, it does not contain a quantitative measure of how often each alternative should be allowed. The profile is a natural extension of the consensus sequence pattern where the observed frequency of individual amino acids is incorporated at each residue position. A further extension is the Hidden Markov model (see *Hidden Markov model* in Section 3.2) where the information of amino acid insertions and deletions is much better represented. While the more flexible data representation causes difficulty for visual comprehension, it generally achieves better prediction accuracy in the computerized motif search. As more sequence data are accumulated and as more sophisticated computational methods are developed, the construction of protein families and the extraction of sequence motifs are automated to maintain useful resources, some of which are shown in Table 2.3 (see *Gene finding and functional predictions* in Section 3.2).

Classification of protein 3D structures

The majority of proteins fall into the categories of globular proteins or membrane proteins. The three-dimensional structures have been resolved for a number of water-soluble globular proteins and for a limited but growing number of membrane proteins. The role of value-added databases is more significant for 3D structural data because the primary database, the Protein Data Bank (Tables 2.1 and 2.2), is simply a repository of all known structures and also because the atomic coordinates data are far more complicated than the sequence data, requiring many different views to be analysed.

The secondary structure elements of α-helices and β-strands have long been recognized as building blocks of protein structures. Globular proteins can be classified into five types depending on the content and arrangement of the secondary structure elements: all α proteins, all β proteins, α/β proteins, α+β proteins, and irregular proteins. The distinction between α/β and α+β is whether α-helices and β-strands are intermixed or separately clustered. Figure 2.13 shows representative proteins in all α, all β, α/β, and α+β categories. This intuitive classification of 3D structures has been refined by computational methods and combined with the sequence similarity classification. The resulting databases (Table 2.3) provide the hierarchy of protein structures. For example, the SCOP database follows the above classification scheme at the top level, and is subdivided into a hierarchy consisting of folds, superfamilies, and families. Generally speaking, 3D structural similarity can be observed in proteins without any significant sequence similarity, i.e. in different superfamilies. This may be the result of either divergent evolution where sequences from the same origin have diverged but still kept structural similarity,

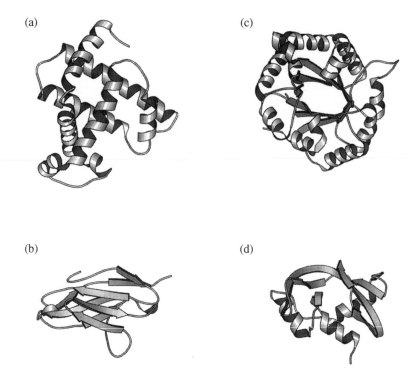

Fig. 2.13. Examples of folds in globular proteins: (a) globin fold in α protein myoglobin (PDB:1MBN), (b) β sandwich in β protein immunoglobulin (PDB:7FAB), (c) TIM barrel in α/β protein triose phosphate isomerase (PDB:1TIM), and (d) a fold in α+β protein ribonuclease A (PDB:7RSA).

or convergent evolution where sequences from different origins have converged into a common structure because of physicochemical constraints and/or functional constraints. The fold represents the level of such structural similarity.

In general the fold is a specific arrangement of the secondary structure elements, which is also called a super-secondary structure or a structural motif. Some folds, such as the α helix bundle, the β sandwich, and the β/α barrel or the TIM barrel, represent overall framework structures that likely reflect physicochemical constraints. Other folds represent specific functional sites, such as the helix-loop-helix EF hand structure for calcium binding and the βαβ Rossman fold for nucleotide binding. These folds may be observed in sequences without global sequence similarity, but usually with local sequence motifs. Figure 2.14 shows the DNA binding motifs of helix-turn-helix, zinc finger, and leucine zipper structures, which are all examples of where a specific function is related to both a structural motif and a sequence motif. This type of sequence–structure–function relationship is an extension of the sequence–function relationship in the motif dictionary. Because of the

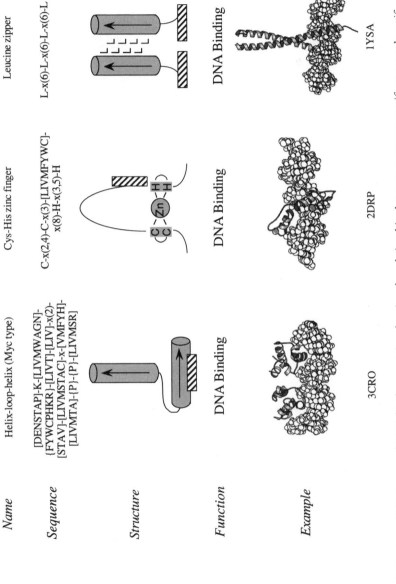

Fig. 2.14. A motif dictionary of DNA binding proteins, showing the relationships between sequence motifs, structural motifs, and functional properties.

limited amount of structural data available, the construction of such an extended motif library has not been undertaken in a systematic way.

Orthologues and paralogues

The term homology is often used to signify the existence of sequence similarity. However, it has been pointed out that homology is ambiguous since it reflects two different mechanisms of evolution that generate similar sequences. Therefore the use of different terms, orthology and paralogy, has been suggested. Orthologous genes share sequence similarity because of speciation from a common ancestor, while paralogous genes are the result of gene duplications within species. Orthologues may imply functional homologues when sequence similarity is high, while paralogues generally have similar but somewhat different functions. In practice, however, the distinction between orthologues and paralogues is not easily

(a)

Organism	epsilon	beta	gamma	alpha	delta	b	c	a
eco	b3731	b3732	b3733	b3734	b3735	b3736	b3737	b3738
bsu	atpC	atpD	atpG	atpA	atpH	atpF	atpE	atpB
mtu	Rv1311	Rv1310	Rv1309	Rv1308	Rv1307	Rv1306	Rv1305	Rv1304
aae	aq_673	aq_2038	aq_2041	aq_679	aq_1588	aq_1586 aq_1587	aq_177	aq_179
syn	slr1330	slr1329	sll1327	sll1326	sll1325	sll1324 sll1323	ssl2615	sll1322

	C	F	A	B	E	K	I	D
bbu			BB0094	BB0093	BB0096	BB0090	BB0091	BB0092
mja	MJ0219	MJ0218	MJ0217	MJ0216	MJ0220 MJ0223	MJ0221	MJ0222	
afu	AF1164	AF1165	AF1166	AF1167	AF1163 AF1158	AF1162 AF1160	AF1159	AF1168

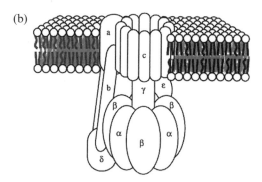

Fig. 2.15. (a) The orthologue group table for F1-F0 ATP synthase (upper) and V-type ATP synthase (lower). (b) An illustration of the three-dimensional structure of F1-F0 ATP synthase.

obtained from sequence information alone, especially when two species have many paralogous genes.

With the availability of complete genome sequences, new efforts have been initiated to compare the entire sets of genes across species and within species. An emerging view is that orthologues and paralogues are better defined for sets of functionally coupled genes rather than for individual genes. A case in point is the molecular assembly. Traditionally, the protein structure formation has been considered as a hierarchy of primary, secondary, tertiary, and quaternary structures. However, the distinction of tertiary and quaternary structures is sometimes mechanical rather than biological; the tertiary structure of a protein with multiple domains could be functionally equivalent to the quaternary structure of the assembly of single-domain subunits. For example, the active transport machinery in bacteria, called the ABC transport system, typically consists of a substrate-binding protein, two membrane proteins for the channel, and two ATP-binding proteins to provide energy. Due to the variety of substances that need be transported across the cell membrane, facultative bacteria contain a large number of paralogues, which are better defined for sets of genes rather than for individual genes. This is because the genes for each transporter are usually encoded in an operon, that is, a stretch of genes in the bacterial genome that is co-transcribed. It appears that the unit of gene duplication is a transcription unit rather than a translation unit. Generally speaking, fused genes or multi-domain proteins are more prevalent in higher organisms, thus, the fusion of genes could be a mechanism to ensure co-transcription when other genes are dispersed in the chromosome.

Of course, biology is not that simple. There are other cases where different combinations of genes are coupled in operon structures and apparently responsible for the formation of molecular machineries with different combinations of subunits. Figure 2.15(a) depicts the gene organization of ATP synthase in bacteria and archaea, which is either of the two types: F1-F0 ATP synthase or V-type ATP synthase. This is an example of the orthologue group table in KEGG (Table 2.3), where each column contains orthologous genes and each row contains a set of genes that forms a functional unit together with the operon information in colour (not shown). By comparing the gene organization with the subunit organization of F1-F0 ATP synthase shown in Fig. 2.15(b), it is interesting to observe the pattern of conserved and diverged subunits in two different types of ATP synthases, although the genes in each group are tightly coupled in an operon structure.

Reactions and interactions

The function is an ambiguous term depending on the level of abstraction that one is interested in. At the level of individual molecules, a protein may be said to be functionally identified when its amino acid sequence is highly similar to those of protein kinases, for example. However, at the level of cellular functions, the function is not identified until the target protein that is phosphorylated is identified, or even until the role of the protein in the biochemical pathway is identified. In essence, the function is an action or a relation that intrinsically requires additional molecules that are acted upon by or related to the original molecule. The majority of the current molecular biology databases treat the function as an attribute of the sequence or structure of an individual molecule. The sequence annotations in the sequence databases or the sequence–function relationships in the motif libraries are based on this concept.

A complementary view is to consider the function as an attribute of an action or a relation between molecules. This is not new in chemical databases, where the concepts of compounds and reactions are equally important. In molecular biology the only reaction data currently well organized are for enzymatic reactions, such as in the LIGAND database (Table 2.3), based on the classification of the EC (Enzyme Commission) numbers. The EC numbering is a hierarchical classification of enzymatic reactions. The hierarchy contains four levels, each designated by a numeral and separated by a comma. At the top level the reactions are classified into six classes:

EC 1. Oxidoreductases
EC 2. Transferases
EC 3. Hydrolases
EC 4. Lyases
EC 5. Isomerases
EC 6. Ligases

It is somewhat confusing that the reactions are named here by the enzymes, but the EC numbering is not a classification of enzyme molecules *per se*. In principle, different molecules with entirely different sequences and 3D structures could be assigned the same EC number as long as the reactions they catalyse are the same. The second and third levels generally specify the chemical groups involved, such as donors and acceptors, and the fourth level designates the substrate specificity.

When the reaction involves small chemical compounds and not macromolecules of gene products, it is possible to organize biological knowledge among the limited possibilities of chemical reactions. In contrast, the modes of interactions among proteins and nucleic acids are virtually limitless when only the physicochemical properties are considered. It is critical to understand biological constraints, such as the repertoire of genes expressed, the cellular and subcellular locations, the developmental stage of the cell, and the substances that surround the cell. In any event, more efforts need be devoted to organize databases of reactions and interactions both in terms of the physicochemical properties and the biological constraints.

Biochemical pathways

The concept of biochemical pathways includes both the metabolic pathways of enzymatic reactions on chemical substances and the regulatory pathways of macro-molecular reactions and interactions. The major portion of the metabolic pathways, called the intermediary metabolism, was mostly clarified by the 1960s and 1970s. When compared to the regulatory pathways, the logic of the metabolic pathways is relatively simple, and knowledge of organic chemistry is directly applicable. Metabolism can be classified into two phases, catabolism where complex molecules are degraded often to generate energy and anabolism in which energy is used to synthesize complex molecules. The exchange of catabolic and anabolic pathways tends to be mediated by a selected number of common metabolites. It has been pointed out that there are 12 precursor metabolites, including pyruvate and acetyl CoA, that serve as hubs in the central metabolism of glycolysis, citrate cycle (TCA cycle), and pentose-phosphate pathway. The intermediary metabolism is a conserved portion of metabolic pathways that is essential to sustain living activities of all organisms. In contrast, the secondary metabolism is a more divergent set of additional pathways, including bacterial biodegradation and biosynthesis of specific compounds. Most of these secondary pathways still need be identified (see *Metabolic network* in Section 4.2).

Metabolism is highly correlated with other cellular activities, such as the regulation of gene expressions and the transport of substances across the cell membrane. All of these activities result from specific interactions among the biological macromolecules of DNAs, RNAs, and proteins. In our definition, the regulatory pathway is a loose categorization of molecular interaction pathways other than the enzymatic reaction pathways, but the distinction is not necessarily clear (see *Protein–protein interaction network* in Section 4.2). For example, the genetic information processing both for transmission and expression is partly macromolecular metabolism but it is also under an ordered network of regulatory mechanisms. The pathways for genetic information transmission include DNA replication, repair, and recombination, while the pathways for genetic information expression include transcipiton, splicing and other post-transcriptional processings, translation, and post-translational modifications. These pathways have mostly been clarified in both prokaryotes and eukaryotes.

Genetic information processing is also regulated at another level of cellular processes, such as the cell cycle for information transmission (DNA replication) and signal transduction for information expression (transcription of specific genes). The cell cycle is an intrinsic property of the cell that leads to cell division, and it has to be accompanied with the correct transmission of genetic information. Signal transduction is the cellular process of activating a set of target genes in response to an external stimulus. The pathway generally starts at the cell surface receptor and ends with the transcription factor, but there can be many variations. The simplest is the bacterial two-component system that consists of only the receptor (sensor kinase) and the transcription factor (response regulator). The cytokine signal transduction pathway includes a third component, while the MAPK signalling and the Ras signalling pathways are much more complex.

When the metabolic pathways became mostly but not necessarily completely known, many researchers moved to the area of signal transduction and other related areas including the cell cycle and apoptosis. Now that these areas are also roughly known, researchers are switching to still more complex processes such as neuronal networks and development. This has been a traditional approach in biochemistry and molecular biology. In contrast, the complete genome sequencing reveals a complete set of genes, which makes it possible in principle to analyse the entire biochemical network that underlies the living activities of a given organism and the environments which it inhabits. However, the current knowledge of biochemical pathways is largely incomplete for such an analysis to be performed successfully. In post-genome informatics it is essential to organize the reference database of biochemical pathways, such as KEGG (Table 2.3), and to develop new computational technologies for pathway comparison, prediction, and reconstruction. More details will be described in Chapter 4.

Genome diversity

The diversity of genomes is twofold: the presence of numerous species on Earth and the polymorphism within each species. Taxonomy is the classification of organisms for the first aspect of genome diversity. The biological domains had traditionally been classified into prokaryotes and eukaryotes, but in the late 1970s the existence of a third domain was identified. Consequently, prokaryotes were divided into bacteria, also known as eubacteria, and the newly identified archaea. Archaea live under extreme environments such as at high temperature (thermophiles) and high salt concentration (halophiles), which may be reminiscent of the early days of Earth. It has since been proposed that, based on molecular sequences, archaea are closer to eukaryotes than bacteria; the tree of life must have first split into bacteria and the common ancestor of archaea and eukaryotes (Fig. 2.16). From a morphological point of view this is counterintuitive, but the machinery for genetic information processing in archaea is in fact more similar to that in eukaryotes, if the similarity is measured by the ribosomal RNA sequences or the subunit organization of RNA polymerases. At the same time, archaea and bacteria are not much different in terms of the genes for metabolism, cellular transport, and cellular architecture. Thus, the archaeal genome appears to be a mosaic of the bacterial and eukaryotic genomes.

The discrepancy between morphological and molecular evidence is also observed when assigning the relationship among animals, plants, and fungi. Common sense suggests that fungi and plants are similar, but according to the analysis of molecular evolution fungi are more similar to animals. The classification of organisms is a difficult and controversial problem even when only the molecular

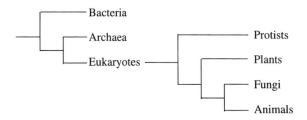

Fig. 2.16. The tree of life showing the relationship of archaea, bacteria, and eukaryotes, as well as the relationship of fungi, plants, and animals.

evidence is considered, since it depends on the genes used to examine evolution and the method used for generating phylogeny. As more complete genome sequences are revealed, it will become possible to compare the entire sets of genes and the resulting biochemical pathways to better define the taxonomy. The most up-to-date consensus view is the NCBI taxonomy database (Table 2.3), which has been adopted by the GenBank and other nucleic acid sequence databases.

The second aspect of genome diversity, the genetic polymorphism within each species, is most important in examining medical consequences of the human genome. The complete genome sequence is a manifestation of an individual. Although a large portion of the genome sequence is conserved within a species or a race, there are polymorphic regions that are naturally present in all genomes and that can be used to fingerprint each individual. The inheritance pattern of polymorphic markers can also be used to identify disease genes in the linkage analysis. OMIM (Online Mendelian Inheritance in Man) is the most comprehensive database for human genes and associated disorders (Table 2.3), and was first released as printed material in 1968 by Victor McKusick. There are many other locus specific databases that collect mutation patterns of specific disorders. It is expected that emerging databases of single nucleotide polymorphisms (SNPs) will play major roles in associating sequence information with heritable phenotypes toward understanding various human disorders as well as the origin and evolution of human species.

3

Sequence analysis of nucleic acids and proteins

3.1 Similarity search

Problems in sequence analysis

DNAs, RNAs, and proteins are linear polymers consisting of repeating nucleotide or amino acid units that assume highly ordered three-dimensional structures to perform specific biological functions. The sequence information on how nucleotides or amino acids are linearly arranged to form a biological macromolecule is relatively easy to obtain by experiments. Thus, the purpose of computerized sequence analysis is to uncover higher structural and functional information encoded in the nucleotide or amino acid sequence data. The topics of sequence analysis are quite extensive, requiring an understanding of both basic algorithms and practical applications, which are being continuously improved upon and expanded. It is not an easy task to summarize sequence analysis in a single chapter, so here we attempt to focus on the basic concepts and to provide an overall picture of sequence analysis.

Computation in biology is different from computation in physics or chemistry. It usually involves processing empirical knowledge acquired from observed data rather than solving first-principle equations, which are virtually non-existent in biology. In fact, the basic idea behind sequence analysis is the empirical knowledge in molecular biology; that is, when two molecules share similar sequences, they are also likely to share similar 3D structures and biological functions because of evolutionary relationships and/or physico-chemical constraints. Thus, the major task of sequence analysis is to find sequence features that can be extended to structural and functional properties. There is a rough distinction of how the data are organized and, consequently, what kinds of computational methods are utilized in sequence analysis, which is categorized in Table 3.1. In the category of similarity search, a basic operation is comparison against each of the known examples stored in a primary database to detect any similarity that can be used for further reasoning. The primary database can be a sequence database or a 3D structural database, and the similarity search involves optimization of a given score function. In the category of knowledge based prediction, known examples are somehow abstracted into empirical rules representing sequence–structure or sequence–function relationships. The rules are stored in computerized forms, such as motif libraries, and are used for practical analysis or for understanding more basic principles. Here a fundamental problem is how to acquire empirical rules, which is related to a

Table 3.1. Search and learning problems in sequence analysis

	Problems in Biological Science		Methods in Computer Science
Similarity search		Pairwise sequence alignment Database search for similar sequences Multiple sequence alignment Phylogenetic tree reconstruction Protein 3D structure alignment	Optimization algorithms – Dynamic programming (DP) – Simulated annealing (SA) – Genetic algorithms (GA) – Hopfield neural network
Structure/function prediction	*ab initio* prediction	RNA secondary structure prediction RNA 3D structure prediction Protein 3D structure prediction	Pattern recognition and learning algorithms – Discriminant analysis – Hierarchical neural network – Hidden Markov model (HMM) – Formal grammar
	Knowledge based prediction	Motif extraction Functional site prediction Cellular localization prediction Coding region prediction Transmembrane segment prediction Protein secondary structure prediction Protein 3D structure prediction	
Molecular classification		Superfamily classification Ortholog/paralog grouping of genes 3D fold classification	Clustering algorithms –Hierarchical cluster analysis – Kohonen neural network

machine learning problem in computer science. The optimization and machine learning methods have been utilized in traditional artificial intelligence applications, such as speech recognition and natural language processing, but they are quite effective in solving problems in biological science as well.

Figure 3.1 illustrates the difference between two approaches for the problem of sequence interpretation, that is how the biological meaning of a newly determined sequence is interpreted using the knowledge of structures and functions in other sequences. In the sequence similarity search or the so-called homology search, a query sequence is compared with each of the database sequences. If any similar sequence is found in the database, and if it is known to be responsible for a specific function, then the query sequence would also have a similar function. The homology search is like searching against all known sentences written in the DNA language, and when matching sentences are found the meaning in the precedents is used to interpret the new sentence. This is not an efficient way of learning a foreign language. Using dictionaries and having knowledge of grammar is more desirable, which is partly accomplished in the motif search. Here empirical knowledge of sequence–function relationships is acquired from the primary databases and organized in the form of a dictionary (see *Protein families and sequence motifs* in Section 2.3). The query sequence is checked with the motif dictionary and if any motif is present it is an indication of a functional site. Unfortunately, it has not been possible to make a comprehensive list of motifs or other empirical rules for all types of functional properties. Thus, in practice, the homology search is still the most preferred approach.

The procedure to find sequence similarity is called sequence alignment. When two sequences are compared it is necessary to make an optimal alignment that best

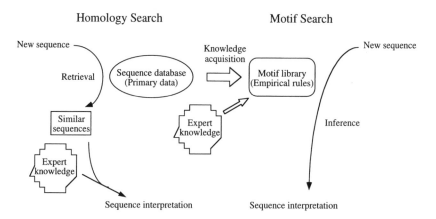

Fig. 3.1. A comparison of the homology search and the motif search for functional interpretation of sequence information.

reveals the similarity relationship of the two. For example, in the alignment of two strings shown below:

ARRS–G

ARKTVG

a gap is inserted in the upper string to maximize the number of matches or to minimize the number of editing operations, such as replacements and insertions/deletions, needed to convert between the two. Although there are three alternative positions to place a gap in this case, it has to be between S and G when these strings represent amino acid sequences. This is because the mismatches of R (arginine) and K (lysine) and of S (serine) and T (threonine) are actually partial matches representing similar amino acid pairs. However, V (valine) in the lower sequence does not have a counterpart; and is likely to be deleted in the upper sequence. In order to incorporate such biological insights, an appropriate measure of similarity has to be defined for each pair of amino acids. The amino acid mutation matrix (see *Amino acid indices* in Section 2.3) provides an empirical scoring scheme for the similarity measure.

Thus, the problem of obtaining the best sequence alignment is equivalent to optimizing a given score function that represents overall similarity, and that is computed from both the similarity measure of amino acids or nucleotides and the penalties for insertions and deletions. There are variations of how the optimal alignment is made. Global alignment compares entire sequences, while local alignment, used in homology searches, detects locally similar regions. Depending on how many sequences are aligned at a time, the distinction of pairwise alignment and multiple alignment is made. Furthermore, the concept of alignment can be extended to other types of data, such as the 3D alignment for comparing two protein 3D structures, the 3D–1D alignment for checking the compatibility of an amino acid sequence against a library of 3D structures (see *Prediction of protein 3D structures* in Section 3.2), and the pathway alignment for comparing two graphs of biochemical pathways (see *Common subgraph* in Section 4.1).

Dynamic programming

The dynamic programming algorithm is a general algorithm for optimization problems. It is also a most fundamental algorithm for understanding the concept of sequence alignment. Suppose that given two strings, AIMS and AMOS, we wish to find the optimal alignment that maximizes the number of matched letters. By inspection, the solution will be:

AIM–S

A–MOS

in which a gap is inserted in each string to obtain the maximum of three matches. In this case, the score function to be optimized is simply the sum of weights at individual positions of the alignment, where the weight is defined one for a match, zero for a mismatch, and zero for an insertion/deletion.

Figure 3.2(a) illustrates the principle of the dynamic programming algorithm. The two strings to be compared are placed on the horizontal and vertical axes of the matrix, which we call the path matrix. Without changing the order of the letters the alignment is made from left to right in both strings, or from upper-left to lower-right in the path matrix. There are three alternative operations at each position of the alignment: to align a letter taken from the horizontal sequence against a letter taken from the vertical sequence, to align a letter taken from the horizontal sequence against a gap in the vertical sequence, and to align a gap in the horizontal sequence against a letter taken from the vertical sequence. These three alternatives are represented by the three possible directions at each node of the path matrix: the diagonal arrow for placing two letters, and the horizontal and vertical arrows for inserting a gap in the vertical and horizontal sequences, respectively. A path starting at the upper-left corner and ending at the lower-right corner of the matrix corresponds to a global alignment of the two strings. Obviously, there are many alternative paths and the problem of finding the optimal sequence alignment has become equivalent to finding the optimal path in the path matrix.

The tree structure shown in Fig. 3.2(b) is an illustration of all possible paths starting at the upper-left corner of the path matrix and resulting from the three-way

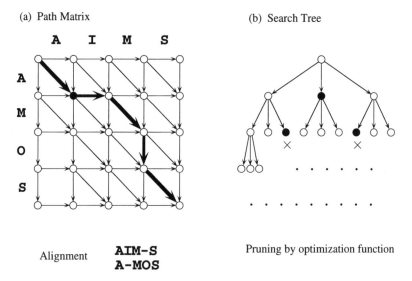

Fig. 3.2. Pairwise sequence alignment by the dynamic programming algorithm. The algorithm involves finding the optimal path in the path matrix (a), which is equivalent to searching the optimal solution in the search tree (b).

branching at each node. The number of branches would become enormous as the sequence lengths become longer and, though not illustrated in the figure, as the number of sequences to be compared increases. Thus, the sequence alignment problem is a typical combinatorial optimization problem in computer science. Any artificial intelligence textbook discusses standard methods to efficiently search such a tree-like structure, including a depth-first search and a width-first search. The sequence alignment requires finding the best solution among all possibilities according to a given criterion, rather than finding any one of the many solutions as in solving a puzzle. The dynamic programming algorithm is a general algorithm to search the tree when it is associated with the score function. The essence of the algorithm can be seen by the filled circles in Fig. 3.2. The three filled circles in the tree (Fig. 3.2(b)) actually represent the same node in the path matrix (Fig. 3.2(a)). The differences are the three possible paths that lead to this location, as well as the scores associated with the paths. If the score is found to be better for the direct diagonal path rather than for the other two multi-step paths consisting of horizontal and vertical steps, then it becomes unnecessary to search all the branches under the two nodes marked with crosses in Fig. 3.2(b), because only the best solution is searched for. Thus, the beauty of the dynamic programming algorithm is that most branches are systematically pruned according to the score function. Because the algorithm effectively evaluates all possibilities, the pairwise sequence alignment problem can be solved rigorously.

Global alignment

Now that we have understood the principle of dynamic programming, let us move on to the actual alignment of biological sequences. The score function to be optimized is the sum of weights at each position of the alignment. The weights are defined by the substitution matrix among four nucleotides or among 20 amino acids and also by the gap penalty. Ideally they should reflect biological processes of mutations and insertions/deletions. For nucleotide sequences, however, the weights are assigned somewhat arbitrarily, usually a fixed value for a match or a mismatch irrespective of the types of base pairs. For the amino acid sequence alignment, which is more powerful since it reveals subtle sequence similarity, the substitution matrix can be constructed from the observed frequency of amino acid mutations adjusted at different degrees of evolutionary divergence. For example, the PAM-250 matrix (Table 2.6) is computed from a set of closely related sequences but adjusted to be applicable to evolutionarily divergent sequences. In the currently available programs, the PAM and BLOSUM series of matrices are most widely used.

Let $w_{s,t}$ be the weight for substituting letter s for letter t or vice versa, which is an element of the symmetric substitution matrix, and let d be the weight for a single letter gap. According to the dynamic programming algorithm, the optimal value of the score function at each node is determined from the three possibilities:

diagonal, horizontal, and vertical paths as shown in Fig. 3.3(a). This is represented by the following equation of the score matrix D:

$$D_{ij} = \max(D_{i-1,j-1} + w_{s(i),t(j)}, D_{i-1,j} + d, D_{i,j-1} + d) \tag{3.1}$$

where $s(i)$ is the i-th letter of the horizontal sequence and $t(j)$ is the j-th letter of the vertical sequence. The initial values of the score matrix are:

$$D_{0,0} = 0$$

$$D_{i,0} = id \text{ (for } i = 1 \text{ to } n)$$

$$D_{0,j} = jd \text{ (for } j = 1 \text{ to } m)$$

where n and m are the lengths of the horizontal and vertical sequences, respectively. Starting at the element $D_{1,1}$, which corresponds to the upper-left corner of the path matrix (Fig. 3.2), and repeatedly applying equation (3.1) to all of the elements, the final value $D_{n,m}$ is the optimal value of the score function. If at each step the optimal path(s) selected from the three possibilities is (are) stored in the path matrix, the overall optimal alignment can be reproduced from the stored connections of the paths by the trace-back procedure starting at the lower-right corner of the path matrix. This procedure to obtain the optimal score and the optimal alignment requires the number of operations to be proportional to the size of the matrix, $n \times m$; thus, this is an $O(n^2)$ algorithm. The original idea of applying the dynamic programming algorithm to sequence alignment was presented by Needleman and Wunsch though in a slightly different form.

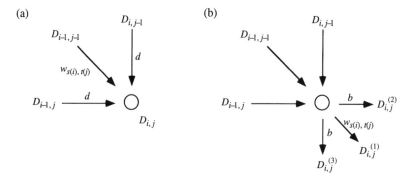

Fig. 3.3. Methods for computing the optimal score in the dynamic programming algorithm. (a) The gap penalty is a constant. (b) The gap penalty is a linear function of the gap length.

When possible molecular mechanisms are considered for insertions and deletions of nucleotides or amino acids, the gap penalty in equation (3.1) is not realistic. A gap of length two is penalized twice that of a gap of length one, which is to
say that it results from a composite of two events. In reality, a gap of consecutive
letters is likely to occur as a single event. By introducing a length-dependent gap
penalty, equation (3.1) is generalized as:

$$D_{i,j} = \max[D_{i-1,j-1} + w_{s(i),t(j)}, \max(D_{i-k,j} + d_k), \max(D_{i,j-k} + d_k)] \qquad (3.2)$$

where d_k is the weight for a gap of length k. This algorithm requires $O(n^3)$ because
of the additional maximization operation in the inner loop. However, when the
length dependence is simplified to consist of an initiation (gap opening) part and
an elongation (gap extension) part, that is, in the form of:

$$a + bk$$

where a and b are constants, the algorithm is still $O(n^2)$ as shown below and in Fig.
3.3(b).

$$D_{i,j}^{(1)} = \max[D_{i-1,j-1}, D_{i-1,j}, D_{i,j-1}] + w_{s(i),t(j)}$$

$$D_{i,j}^{(2)} = \max[(D_{i-1,j-1} + a), D_{i-1,j}, (D_{i,j-1} + a)] + b \qquad (3.3)$$

$$D_{i,j}^{(3)} = \max[(D_{i-1,j-1} + a), (D_{i-1,j} + a), D_{i,j-1}] + b$$

Note that the scores for the horizontal and vertical directions are stored and compared separately, distinguishing between the initiation and the elongation of a gap
in the respective direction. The score for the diagonal direction is the same as in
equation (3.1).

Local alignment

The optimal path that extends from the upper-left corner to the lower-right corner of the path matrix (Fig. 3.2) is a global alignment. Now let us consider how to
obtain local alignments, which correspond to shorter, localized paths in the path
matrix. Figure 3.4 illustrates the notion of local optimality. The global alignment
was obtained from equation (3.1), with the initial value of zero for only one element, $D_{0,0} = 0$, of the score matrix (Fig. 3.4(a)); that is, the first letter of the horizontal sequence and the first letter of the vertical sequence were the starting point
of the alignment. In contrast, as shown in Fig. 3.4(b), by assigning $D_{0,j} = 0$ for all
values of $j = 1$ to m, any letter within the horizontal sequence could be a starting
point without any penalty. This is useful for detecting multiple matches within a
horizontal sequence consisting of multiple domains, each of which is similar to the

vertical sequence. The global alignment procedure would detect only one of them. Thus, in this case the vertical sequence is globally aligned but the horizontal sequence is locally aligned.

By extending this notion, the local alignment of both sequences might possibly be obtained by assigning $D_{i,0} = 0$ for all values of $i = 1$ to n and $D_{0,j} = 0$ for all values of $j = 1$ to m. Any letter within either sequence can now be a starting point of alignment without any penalty. As shown in Fig. 3.4(c), Sellers proposed that the dynamic programming procedure should be applied twice in both directions, starting at the upper-left corner and at the lower-right corner, and the logical product of the two path matrices generated should be taken to identify the optimal local alignment. In practice, however, this procedure was not very successful. It was necessary somehow to incorporate the zero values inside the matrix rather than just use those at the edges of the matrix.

Suppose that the elements $w_{s,t}$ of the substitution matrix take positive and negative values depending on favourable and unfavourable substitutions, respectively, and that the gap penalty is negative because gaps are always unfavourable. By considering zero as a neutral value, equation (3.1) is modified as:

$$D_{i,j} = \max(D_{i-1,j-1} + w_{s(i),t(j)}, D_{i-1,j} + d, D_{i,j-1} + d, 0) \tag{3.4}$$

which is equivalent to forcing the score value to be zero whenever it becomes negative. When this happens the optimal path is not entered in the path matrix, which effectively separates out local clusters of favourable regions. A match of letters with a positive score can thus either extend an existing cluster or initiate a new cluster

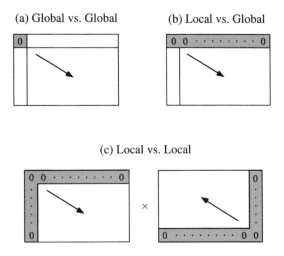

(a) Global vs. Global   (b) Local vs. Global

(c) Local vs. Local

Fig. 3.4. Concepts of global and local optimality in the pairwise sequence alignment. The distinction is made as to how the initial values are assigned to the path matrix.

with the initial score of zero, i.e. without inheriting negative values of unfavourable regions. The Smith–Waterman algorithm is based on this notion of local optimality and makes the best local alignment by the trace-back procedure starting at the matrix element with the maximum alignment score. The Goad–Kanehisa algorithm additionally incorporates the Sellers procedure of taking the logical product of two path matrices and finds all local alignments above a given threshold.

Database search

The database search for similar sequences, also known as the homology search, uses the local alignment procedure to compare the query sequence with each of the database sequences. When the alignment is done by the dynamic programming algorithm, which is $O(n^2)$, an enormous number of operations is required to compare a sequence of reasonable length against the current sequence database. One way to perform an efficient search is to use a specialized machine.

The dynamic programming can be made efficient by parallel processing. A

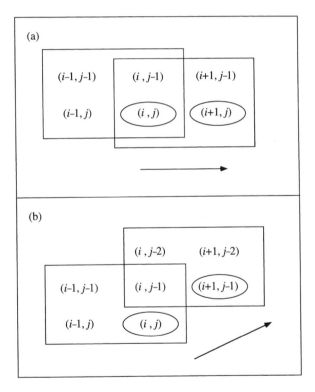

Fig. 3.5. The order of computing matrix elements in the path matrix, which is suitable for (a) sequential processing and (b) parallel processing.

standard way of computing the score matrix element $D_{i,j}$ is in the row-wise direction shown in Fig. 3.5(a), where each element is computed from left to right in each row, taken from top to bottom in the matrix. In order to compute the element $D_{i+1,j}$, the previous element $D_{i,j}$ is necessary according to equation (3.1) or (3.4). This is a sequential processing where the computation of $D_{i+1,j}$ cannot be initiated until the computation of $D_{i,j}$ is completed. In contrast, if the order of computation is taken in the anti-diagonal direction as shown in Fig. 3.5(b), the computation of $D_{i,j}$ and the computation of $D_{i+1,j-1}$ can be performed simultaneously because all necessary elements are pre-computed. This parallel dynamic programming algorithm can be implemented in a computer with special hardware.

Another way of parallelization is to split the database into smaller pieces where the database search jobs are assigned to multiple nodes of a parallel machine or to a cluster of workstations. In this case, the increase in speed is directly proportional to the number of nodes. In contrast, the parallelization of the dynamic programming algorithm is not very efficient. This is because the length of the anti-diagonal varies from one to the maximum length depending on its location in the matrix (Fig. 3.5(b)).

Although dynamic programming is time-consuming, it finds the most rigorous solution and is most sensitive for detecting subtle similarities. It is the ultimate method for comparing amino acid sequences, because the score function reflects similarities between amino acids and, above all, the database size is relatively small. However, it may not be worthwhile utilizing dynamic programming in the database search of nucleotide sequences, which is usually just a string match of identical nucleotides. Let us now examine two popular algorithms, FASTA and BLAST, which are approximate, but more efficient algorithms for the database search.

FASTA algorithm

The dot matrix is a primitive, visual way of identifying locally similar regions of two sequences. The matrix is formed by horizontally and vertically placed sequences and by dots that are entered for all positions corresponding to matching pairs of letters in the two sequences. By visual inspection local similarities can be detected as diagonal stretches of consecutive dots, or clusters of diagonal stretches when mismatches and small insertions/deletions are allowed. As illustrated in Fig. 3.6, the area occupied by such diagonal stretches is minuscule in comparison to the entire area of the matrix. Practically speaking, the dynamic programming algorithm searches the entire matrix and thus spends a great deal of time performing unproductive computations in order to avoid the slim risk of missing any similarity.

The FASTA method was invented to incorporate this characteristic property of the dot matrix. It is designed to limit the search area that the dynamic programming will examine. Suppose that a diagonal stretch of consecutive dots can be found rapidly by a certain method. In Fig. 3.6 the position of a diagonal is

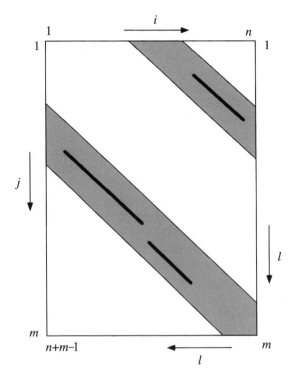

Fig. 3.6. The dynamic programming algorithm can be applied to limited areas, rather than to the entire matrix, after rapidly searching the diagonals that contain candidate matches.

specified by l whose value ranges from 1 to $n + m - 1$ where n and m are the lengths of the sequences being compared. By considering the possibility of insertions and deletions that cause shifts of diagonals, limited areas (shaded in Fig. 3–6) are defined around the identified diagonals and used for the dynamic programming search. This procedure significantly reduces the area to be searched, and thus the computation time needed by the dynamic programming algorithm.

In FASTA the initial search of candidate diagonals is done by hashing, which is a scheme for rapid access to data items using a special look-up table called a hash table. Figure 3.7 illustrates the principle of hashing where the query sequence is hashed according to the single letters A, C, G, and T. The dot matrix indicates the locations of nucleotide matches when the horizontal query sequence is compared with the vertical database sequence. A naive $O(n^2)$ algorithm to identify these matches is to inspect each row in turn and compare a letter in the database sequence against all letters in the query sequence. Since the query sequence is used over and over again, it would be more efficient to somehow preprocess the data. The hash table in Fig. 3.7 represents such preprocessing. The data is classified according to the key, in this case the four single letters, and the table contains

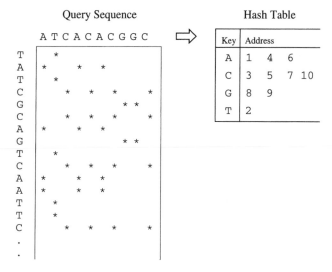

Fig. 3.7. The hashing technique for rapid sequence comparison. In this case the horizontal sequence is converted to a hash table, which contains the locations of the four nucleotides.

pointers to the original data, in this case the positions of each nucleotide in the query sequence. To enter dots in a row of the dot matrix, it is now sufficient to look up only the corresponding key in the hash table. For example, for nucleotide A in the database sequence, dots should be entered at the first, fourth, and sixth elements in the row. The number of operations required is now proportional to the mean row size of the hash table, which is on average one fourth of the query sequence length when the composition of the four nucleotides is even. Therefore, hashing by single letters makes the search four times faster.

Hashing can also be done by double, triple, or generally k-tuple letters. The tuple size of $k = 2$ and 3 would make the search 16 times and 64 times faster, respectively, for nucleotide sequences. However, when doublets are matched, for example, AC and AG are considered a mismatch rather than a partial match of one of the two letters. Therefore, as the tuple size increases the sensitivity will be reduced in exchange for the speedup. In the FASTA program the default parameter value for the tuple size is six for nucleotide sequences and two for amino acid sequences.

BLAST algorithm

The dynamic programming algorithm for pairwise sequence alignment can be written in a simple set of codes in a computer program, which specifies instructions to examine virtually all combinations and to compute the rigorous solution. In contrast, when humans try to align two sequences they do not examine all possibilities but instead resort to a more intuitive method by considering odds to reach the solution faster. Perhaps, they first search for a block of letters that commonly appears in both sequences and then try to extend the matched region. In fact, this makes sense biologically because two similar sequences tend to share cores of well-conserved segments that are often represented as sequence motifs. BLAST is a heuristic algorithm, meaning that it incorporates good guesses based on the knowledge of how sequences are related.

The BLAST algorithm also has a sound mathematical foundation concerning the statistics of local alignment scores in random sequences. The optimal ungapped alignment between two sequences is called a maximal-scoring segment pair (MSP). The score distribution of MSPs in the limit of long sequences with a random sequence model is known to follow an extreme value distribution. In practice, this statistic is extrapolated to assess the significance of a high-scoring segment pair (HSP), which is a locally maximal segment above a given threshold score. The expected frequency E of observing an HSP with score S or higher in the comparison of two random sequences of lengths n and m is given by:

$$E = Knm \exp(-\lambda S) \qquad (3.5)$$

where K and λ are Karlin–Altschul parameters.

The BLAST algorithm is based on a systematic search of conserved words. A word is a tuple of letters where the default lengths are $W = 3$ for amino acids and $W = 11$ for nucleotides. First, the query sequence is decomposed into words of length W and a list of these words and similar words is created. Similar words are collected from all other combinations of W-tuples when the score for ungapped alignment with the original word exceeds a given threshold T. If the self-comparison of a word in the query sequence gives a score blow T, it is removed from the list. Thus, when any word in the list is found in the database sequence, it can become a core segment used to start a local alignment.

In order to search all occurrences of all words in the database sequence, the finite state automaton can be used. The finite state automaton is an abstract machine consisting of sets of states and transitions, as well as inputs and outputs. In this case, all words in the list are encoded by the states and transitions of an automaton, and when the database sequence is given as an input the automaton produces an output of all hits. Figure 3.8 illustrates an example where three triplet patterns, ABA, BAA, and BAB, are searched for in a sequence consisting of three letters, A, B, and C. Starting from the initial state Q_0, the transition will occur depending on what letter is encountered next in the sequence. The transitions marked by thick arrows are associated with outputs indicating that the patterns have been found. Therefore, reading through the sequence just once, all occurrences of all patterns can be found.

Once similar words are found in the database sequence, BLAST will make an ungapped local alignment of the query sequence and the database sequence, starting at each of the words found and extending on both sides as long as the score

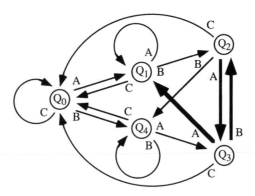

Fig. 3.8. An example of the finite state automaton for pattern matching.

increases. When the score of the longest segment exceeds the threshold S, the segment is considered an HSP whose statistical significance can be evaluated by equation (3.5). When multiple HSPs are found the statistical significance is examined for their combination. There have been many improvements on the basic BLAST algorithm outlined above, including the gapped alignment and the position-specific scoring scheme.

Statistical significance

A major concern when interpreting the database search result is whether the similarity found is biologically significant, especially when the similarity is only marginal. Because good alignments can occur by chance alone, the statistics of alignment scores help to assess the significance. In the BLAST program the E value, equation (3.5), for ungapped local alignments in two sequences is extended to compute the expected frequency in the entire database search. In FASTA, however, such computation is not possible. According to equation (3.5) the average score S for a query sequence with fixed length n increases with the logarithm of the length m for the database sequence. Thus, the distribution of sequence lengths in the database can be used to estimate empirically the E value in the FASTA program.

Another type of statistic, the Z value, is also utilized. It is based on an empirical score distribution in randomized sequences. The Z value can be computed as follows. Suppose that the observed score is S for the global (or best local) alignment of two sequences. By repeating k times the procedures of randomizing each sequence and making an optimal alignment, a series of scores $s_1, s_2, \ldots, s_k$ can be obtained. With the mean μ and the standard deviation σ, the Z value is defined as:

$$Z = \frac{S - \mu}{\sigma} \tag{3.6}$$

which is the difference from the mean, measured by the standard deviation units. If the score distribution in randomized sequences is normal, Z values of 4 and 5 correspond to probabilities of 3×10^{-5} and 3×10^{-6}, respectively. In reality, however, the distribution is not normal; equation (3.5) suggests that the extreme value distribution decays exponentially in S rather than S^2 as in a normal distribution. Thus, a higher Z value should be taken as a threshold for significant similarity.

When the Z value is computed, random sequences are generated preserving the composition of amino acids or nucleotides. In fact, the compositional bias can easily exaggerate the statistical significance. For example, a stretch of just A and T in the query nucleotide sequence will match many AT rich regions in the database sequences. Therefore, before the database search is performed, filtering programs are sometimes used to mask regions of restricted compositions of amino acids or nucleotides, called low-complexity regions, in the query sequence.

Multiple alignment

A weak similarity in pairwise sequence comparison may still turn out to be a strong feature when a number of sequences are collected and compared at one time. The multiple sequence alignment is a simultaneous comparison of a group of sequences, and is extremely useful in identifying locally conserved regions of functional and/or structural importance. The resulting sequence features, which may be abstracted in consensus patterns, blocks, or profiles (see Fig. 2.12), can then be used to detect subtle similarity. The multiple sequence alignment also helps to examine global similarity and evolutionary relationships within a sequence family. These problems can be solved by generalizing the optimization algorithms for local and global pairwise alignments. The score function to be optimized is computed from the sum of all possible pairs in a group of sequences, although alternative definitions can also be made.

Conceptually, the pairwise dynamic programming algorithm can easily be extended to the problem of multiple sequence alignment. The two-dimensional matrix (Fig. 3.2) would simply be expanded to an n-dimensional matrix. In practice, the simultaneous alignment of three sequences is readily available, and when only a limited portion of the n-dimensional space is searched, as in Fig. 3.6 for two dimensions, the multiple alignment of seven or eight protein sequences is manageable. Beyond this, however, the combinatorics would explode and the rigorous algorithm becomes inaccessible by current computers both in terms of memory space and computation time. It is necessary to resort to an approximate algorithm of combinatorial optimization.

Here we introduce heuristic methods for the global multiple sequence alignment. The simplest one is to somehow merge the results of pairwise alignments to obtain a multiple alignment. Suppose that there are three sequences to be compared, A, B, and C. When the optimal pairwise alignments A–B and B–C are combined, the alignment of A and C is automatically determined, but it is in general different from the optimal alignment A–C because gaps are likely to be inserted in different positions. To minimize such discrepancies, it is better to select pairs with the best scores or the smallest number of gaps first. However, this type of simple combination does not work well.

An improvement is a tree-based progressive alignment method shown in Fig. 3.9. First, the tree is constructed by the hierarchical cluster analysis of a set of sequences. Cluster analysis is a type of multivariate analysis used to group similar data given an appropriate measure of distance. In this example the distance between two sequences is defined by the optimal pairwise alignment score that is computed by either the dynamic programming or another algorithm. The distance to a group (cluster) of sequences can be measured in many ways. Single linkage is to measure the distance according to the nearest point in a cluster, complete linkage is according to the furthest point in a cluster, and there are different types of mean distances computed from a collection of points in a cluster. The result of clustering can be visualized by a tree-like structure called a dendrogram (Fig. 3.9(a)), in which similarities and linkages are hierarchically represented. According to this dendrogram, a progressive multiple sequence alignment is performed by combining the alignments between two sequences, between a sequence and a group of sequences, and between two groups of sequences. As indicated in Fig. 3.9(b), the pairwise group alignment involves a matrix made up of a group of sequences placed horizontally and another group of sequences placed vertically. The problem here is to find an optimal alignment of two groups without changing the predetermined alignment within each group. The score function to be optimized can be computed as the sum of pairs formed by sequences from the two groups.

The tree-based progressive alignment can further be fine-tuned for practical implementation by adjusting similarity weights and gap penalties that depend on the progress in the tree or the position in the alignment. However, once an alignment is made, it cannot be modified in a later stage of the group alignment. To overcome this obstacle an iterative procedure can be introduced to successively improve the overall multiple alignment. Starting from a set of sequences that are unaligned or partially aligned, the set is randomly divided into two groups and the optimal group alignment is then performed to produce a better multiple alignment. The steps of random division and pairwise group alignment are repeated a number of times until the score function converges to a predefined range of values. The iterative improvement method is time-consuming, but in practice it probably provides the best solution.

Phylogenetic analysis

The presence or absence of specific characters in molecular, physiological, morphological, and/or behavioural data can be used to infer evolutionary trees, also known as phylogenies, among a set of organism groups, or taxa. The amino acid or nucleotide sequence data are especially suited for defining such characters and are

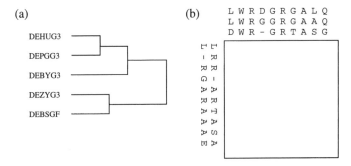

Fig. 3.9. The tree-based progressive method for multiple sequence alignment, which utilizes: (a) a dendrogram obtained by cluster analysis and (b) group alignment for pairwise comparison of groups of sequences.

widely used in phylogenetic analysis. It is beyond the scope of this book to review all the different methods for phylogenetic tree reconstruction from sequence data. Here we will only briefly look at the distance matrix approach and the parsimony approach. In both approaches a proper multiple sequence alignment is a prerequisite of tree reconstruction.

The topology representing the similarity relationship of three sequences A, B, and C is unique, as shown in Fig. 3.10(a). The actual lengths of the edges (branches) from the three exterior nodes (extant sequences) to an interior node (common ancestor) in this unrooted tree can be determined from the three pairwise distances AB, AC, and BC. When the fourth sequence D is added, there are three alternative positions to place an additional interior node in the tree, giving rise to three possible topologies for the four sequences as shown in Fig. 3.10(b). The number of pairwise distances is six in this case, AB, AC, AD, BC, BD, and CD, but there are only five branches that are adjustable. In general, there are $n(n-1)/2$ pairwise distances for a set of n sequences. The number of tree topologies is $3.5 \ldots (2n-5)$ and each tree consists of n exterior nodes, $n-2$ interior nodes, and $2n-3$ branches. Therefore, the problem of reconstructing the tree that best represents $n(n-1)/2$ pairwise distances is an optimization problem.

A rapid method for finding an approximation to the optimal tree is the neighbour-joining method of Saitou and Nei. First, the pairwise distances are computed and stored in the distance matrix for a set of aligned sequences. Starting from the star topology, sequences are successively clustered by joining neighbours in a pairwise fashion; hence the size of the distance matrix is reduced one by one. The pair to be clustered at each step is selected from all possible pairs according to the smallest sum of the branch lengths for the resulting tree. In contrast, UPGMA (unweighted pair-group method by arithmetic averaging) clusters sequences according to the smallest distance in the distance matrix.

While the distance matrix approach is based on the overall similarity of sequences, the parsimony approach partitions similarities on a character-by-character

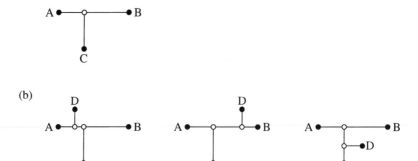

Fig. 3.10. Possible tree topologies in the phylogenetic analysis of: (a) three sequences or (b) four sequences. Filled circles represent extant sequences, while open circles represent common ancestors.

basis. In the multiple alignment of nucleotide or amino acid sequences each column represents a character, the presence or absence of a substitution or an insertion/deletion event. The principle of parsimony is the principle of minimum evolution; namely, the tree is reconstructed to account for the fewest possible mutational events. The parsimony analysis requires at least four taxa, where there are three unrooted tree topologies as shown in Fig. 3.10(b). The number of mutational events can be counted for each tree according to the multiple alignment, and the best tree is selected by identifying the one with the minimum number.

Although the concept of phylogeny contains evolutionary history and development from ancestors, any algorithm for phylogenetic reconstruction provides only an unrooted tree. In order to determine the ancestral node, or to root a tree, knowledge of an outgroup, or a taxon outside of the analysed group, is usually required. In any tree, small variations of input data leave some branches untouched while bringing changes to others. Bootstrapping is a statistical method commonly used to estimate the reproducibility of individual branches.

Simulated annealing

Let us now consider the optimization problem from a somewhat different perspective. Figure 3.11 is a schematic illustration of the optimization function E plotted against the conceptual variable space x. For the problem of multiple sequence alignment, the abscissa x is the space representing all possible alignments and the ordinate E is the negative score function, which we call the energy function. Thus, the optimization here involves minimization, rather than maximization, of the energy function. When the overall energy landscape is known, as in this figure, it is obvious where the global minimum is located. However, in multiple sequence alignment of many sequences the combinatorics becomes too enormous to compute all

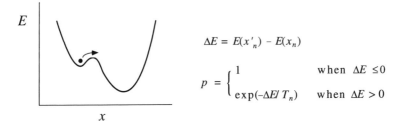

$$\Delta E = E(x'_n) - E(x_n)$$

$$p = \begin{cases} 1 & \text{when } \Delta E \leq 0 \\ \exp(-\Delta E/ T_n) & \text{when } \Delta E > 0 \end{cases}$$

Fig. 3.11. Simulated annealing and Metropolis Monte Carlo methods are based on the concept of thermal fluctuations in the energy function.

possibilities. It is therefore difficult to tell whether the minimum found by an approximate method is only a local minimum or the real global minimum. Generally speaking, the local minima problem is inherent in any large-scale optimization problem in which rigorous computation is impossible.

Simulated annealing (SA) is a stochastic method to search for the global minimum without being trapped in local minima, and it is generally applicable to many optimization problems, including the multiple sequence alignment. Annealing is the process of heating and slowly cooling metal or other materials. Slow cooling is required to obtain tough metal in the most stable state, or the global minimum, of the crystal structure, while rapid cooling is likely to produce brittle metal which is considered to be in a meta-stable state, or a local minimum, of the crystal structure. The energy function shown in Fig. 3.11 contains a local minimum in addition to the global minimum, and the principle of annealing is illustrated by the arrow escaping from the local minimum. This energy change is caused by thermal fluctuations, which are larger at higher temperatures.

If humans tried to align many sequences at a time, they would introduce a small modification, such as a shift or a gap insertion, to the current alignment, evaluate the outcome, and either accept or reject the modification. They would repeat this cycle a number of times until they are satisfied with the overall alignment. These steps can be implemented in a simulated annealing algorithm. Starting with a given alignment of a set of sequences, a small modification is repeatedly introduced to the current alignment. This is done randomly, which of course is different from human alignment with heuristics. When the modification is favourable in terms of the given score (energy) function, it is accepted. However, if the modification is rejected whenever the score function becomes worse, it is impossible to escape from a local minimum. Thus, an unfavourable modification is also accepted according to the following probability

$$p = \exp(-\Delta E/ T)$$

where ΔE is the increment value of the energy function and T is the parameter that plays the role of the temperature. This probability represents the Boltzman factor of the thermal equilibrium distribution function, and it was first implemented in the Monte Carlo method of computer simulation by Metropolis. In simulated annealing, the Metropolis Monte Carlo procedure is repeated at gradually decreasing temperatures. At each temperature the Metropolis Monte Carlo procedure is performed until the thermal equilibrium condition is virtually obtained, and then the temperature is lowered by a small amount.

Though simulated annealing is a probabilistic simulation method, it has a mathematical foundation. It will find the global minimum with the limits that the number of Metropolis Monte Carlo steps is infinite and that the decrement of the temperature is infinitesimal. This of course is practically impossible, and success thus depends on the cooling schedule that includes the initial (highest) and final (lowest) temperatures, the decrements of temperature, and the number of Monte Carlo steps.

Genetic algorithms

In order to conform to the mathematical foundation, each modification in simulated annealing must be small enough to satisfy the so-called detailed balance condition. In practice, however, larger modifications are introduced to speed up convergence to the equilibrium state. Genetic algorithms, which form another class of stochastic solutions to the optimization problem, more actively incorporate drastic changes as well as small modifications. A drastic change is called a crossover and a small change is called a mutation, based on a superficial resemblance to the actual genetic processes in biology. While simulated annealing has a sound basis from the real physical phenomenon, genetic algorithms do not have much relevance to real biology. Therefore, the terms in genetic algorithms should not be associated with any biological meanings.

In genetic algorithms, an individual (solution) with the highest fitness (optimization function) is searched for among a population of individuals through the process of simulated evolution. An individual is represented by the chromosome, which is in essence a set of character strings or bit strings. Suppose an optimal three-dimensional RNA structure is searched for among many possible conformations. The conformation of a single nucleotide in RNA can be specified by seven variables in the dihedral angle representation. If each variable is assumed to take n bits with discrete values, the entire RNA conformation with k nucleotides is represented by the chromosome of $k \times 7n$ bits. Here the bit string is encoded in Gray code, which is often used in genetic algorithms for mapping between a decimal number and a bit string. Mapping each digit of a decimal number to a string of four bits corresponds to choosing 10 strings from 16 possibilities. In Gray code, two neighbouring decimal digits are always represented by adjacent bit strings that differ by only one bit position. In the standard binary code, for example, the decimal digits 3 and 4 are, respectively, 0011 and 0100, which differ by three bit positions.

The actual procedure of simulated evolution can be done in many different ways. An example is the following. The initial population consists of N individuals. First, N chromosomes are copied and subject to mutation operations. A mutation is introduced at a random bit position of a randomly selected chromosome. After the mutation operation is performed for a predefined number of times, the two populations with and without mutations are mixed and subject to crossover operations. A crossover here is an operation to exchange parts of two randomly selected chromosomes at a random crossover point, and the fittest of the two new chromosomes is kept in the population. After the crossover operation is repeated M times, the total population becomes $2N + M$. According to the selection probability that is determined from the fitness function, N individuals are selected from this population to start another cycle of mutation and crossover operations. The cycle is repeated until a solution with reasonable fitness is obtained.

3.2   Prediction of structures and functions

Thermodynamic principle

The amino acid sequence is usually considered to contain all the necessary information to fold a protein molecule into its native 3D structure under the physiological conditions. By changing the environmental variables such as temperature, pressure, and solvent conditions, the protein undergoes a structural transition from the native state to the denatured (unfolded) state. However, once the variables are returned to the physiological conditions, the protein refolds spontaneously and regains its activity. This thermodynamic principle was first established in the 1950s by Christian Anfinsen's denaturation–renaturation experiments on ribonuclease and later confirmed by similar experiments on other small globular proteins.

If protein structure formation is thermodynamically determined, then it should be possible to predict the 3D structure computationally by searching for the most stable structure for a given amino acid sequence. This may be done by defining a conformational energy function for the protein 3D structure and then minimizing the function. This type of *ab initio* prediction (Table 3.1), or a prediction based on first principles, has not been very successful because of the difficulty of defining an appropriate conformational energy function that encompasses the folded state and the unfolded state, and also because of the local minima problem associated with any optimization procedure. In contrast, knowledge based prediction of 3D structures has been far more successful and is of practical use. It utilizes knowledge of actual protein 3D structures determined by X-ray crystallography and NMR experiments, and establishes empirical sequence–structure relationships. Since there has been little formulation of first principles in biology, the knowledge based prediction is the main approach to most structure/function prediction problems of proteins and nucleic acids (Table 3.1).

Although the conformational energy calculation may not be practical for structure prediction from the entirely unfolded amino acid chain, it is useful for predicting 3D structure perturbations around the known native structure. The conformational energy E may be represented in the following form:

$$E = E_b + E_\theta + E_\phi + E_{vdw} + E_{el}$$

The first two terms are contributions from the bond lengths and the bond angles of covalent bonds:

$$E_b = \Sigma K_b(r - r_0)^2$$

$$E_\theta = \Sigma K_\theta(\theta - \theta_0)^2$$

The third term is the potential for the dihedral angles that specify the backbone conformation:

$$E_\phi = \Sigma K_\phi[1 + \cos(n\phi - \delta)]$$

The last two terms represent van der Waals interactions and electrostatic interactions of noncovalent bonds:

$$E_{vdw} = \sum_{i<j} \left[\frac{A_{ij}}{r_{ij}^{12}} - \frac{B_{ij}}{r_{ij}^6} \right]$$

$$E_{el} = \sum_{i<j} \frac{q_i q_j}{\varepsilon r_{ij}}$$

Since the parameters in these equations are determined empirically, the conformational energy is not really *ab initio* at the level of quantum chemistry. The conformational energy calculation has been utilized in protein engineering. In order to design, for example, a new protein with improved enzyme activity or modified substrate specificity, predictions can be made concerning which amino acids are to be replaced by site-directed mutagenesis in the known protein structure.

Prediction of RNA secondary structures

The secondary structure of a single-stranded RNA molecule is stabilized by localized regions of self-complementary base pairs (see Fig. 1.8(a)). A stem is a double-helical base-paired region and a loop is a single-stranded region between stems. There are different types of loops—a hairpin loop, an internal loop, a bulge loop, and a branch loop, as illustrated in Fig. 3.12. The prediction of RNA secondary structure is a type of *ab initio* prediction, which involves minimization of an energy function that contains contributions from both stems and different types of loops. The prediction of RNA secondary structure is also a type of sequence alignment problem, which compares a single sequence against itself and utilizes similarity scores for stems and gap penalties for loops. Thus, the optimal RNA secondary structure can be found by the dynamic programming algorithm for the energy function $f_{i,j}$:

$$f_{i,j} = \min[f_{i+1,j-1} + \alpha_{i,j}, \min(f_{i+k,j} + \beta_k), \min(f_{i,j-k} + \beta_k), \min(f_{i+k,j-l} + \gamma_{k+l}),$$
$$\min(f_{i+k,j'} + f_{i',j-l} + \varepsilon_{k+l+j'-i}), \delta_{j-i}]$$

Here the parameter $\alpha_{i,j}$ represents the energy contribution from base pair i, j and the preceding pair in the stem region. This is assigned to a doublet of successive base pairs, rather than a single base pair, in order to incorporate the base-stacking energy as well as the base-pairing energy. In addition to the standard Watson–Crick base pairs, A–U and G–C, the pair G–U is also considered. The parameters β_k, γ_k, ε_k, and δ_k are, respectively, for bulge, internal, branch, and hairpin loops of length k. The bulge loop has two possibilities depending on the strand in which the loop exists. This algorithm requires the number of operations to be in the order of $O(n^3)$, which is similar to the sequence alignment with a generalized length-dependent gap penalty (equation 3.2).

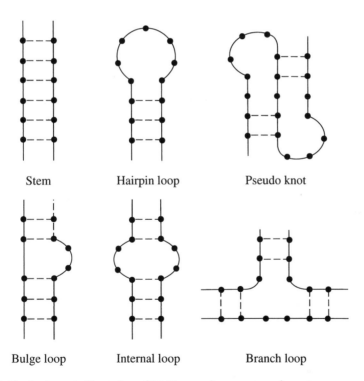

| Stem | Hairpin loop | Pseudo knot |

| Bulge loop | Internal loop | Branch loop |

Fig. 3.12. A schematic illustration of RNA secondary structure elements.

The parameters for the energy contributions are determined empirically, as in the case of protein conformational energy. The base-stacking and base-pairing energies have been measured from thermodynamic experiments of synthetic oligonucleotides, but the contributions of loops are mostly estimated by theoretical considerations. The dynamic programming algorithm is a rigorous algorithm that guarantees to find the global minimum; there is no local minima problem here. However, it is still possible that a small variation of a parameter will produce a very different overall structure and that there may be many other solutions that are almost as good as the optimal solution. For the latter possibility the dynamic programming algorithm can be extended to find all sub-optimal solutions within a given range. For the former possibility it is desirable to somehow identify groups of structures with different arrangements of secondary structure elements.

Hopfield neural network

An alternative approach to RNA secondary structure prediction is just to consider combinations of possible stem regions, which is enough to identify different overall structures. This can be done first by enumerating all possible stems (complementary base-paired regions) greater than a given length, and then by finding the optimal combinations with compatible (non-overlapping) stems according to a given energy function. Although this approach generally disregards correct evaluation of loops, the energy function may not be affected much because most contributions come from the stems anyway. Note that since two stems can be made compatible by unwinding the ends of the stems, all possible sub-stems within a stem should be considered as candidates for combination. Therefore, it is necessary to examine a huge number of possibilities, which is again a combinatorial optimization problem. Let us consider a unique way of solving this problem.

Artificial neural networks (ANNs) form a class of computational methods that can be applied to a diverse range of problems including machine learning (see *Hierarchical neural network* below) and classification. In fact, an ANN is better defined as a computational paradigm that is motivated by the biological neural network and its processing power in the human brain. The Hopfield neural network was designed as a simple model for associative memory in analogy to the spin-glass system in physics. It is a symmetrically connected network where every node (neuron) is connected to every other node. It has the energy function E in the quadratic form:

$$E = \tfrac{1}{2} \sum_{i=1}^{n} \sum_{j=1}^{n} w_{ij} v_i v_j - \sum_{i=1}^{n} \theta_i v_i$$

where v_i is the binary variable for each node, which takes the value of either 1 or 0, and w_{ij} and θ_i are parameters. The minimization of this energy function is done by a stochastic procedure. By randomly selecting a node its state is modified according to the input–output relation of a neural network; in this case, according to the states of the surrounding nodes together with an activation function:

$$u_i = \sum_{j=1}^{n} w_{ij}v_j + \theta_i$$

$$v_i = \frac{1}{1 + \exp(-u_i/T)}$$

Here a technique similar to that of simulated annealing can be used with the temperature parameter T, as a better way of finding the global minimum.

For the problem of RNA secondary structure prediction, the energy function may be written as:

$$E = \sum_{i=1}^{n} e_i v_i + \lambda \frac{\max(|e_i|)}{2} \sum_{i=1}^{n} \sum_{j=1}^{n} c_{ij} v_i v_j$$

where e_i is the energy for stem i, v_i is 1 or 0 depending on whether stem i is selected or not, and c_{ij} is 1 or 0 depending on whether stems i and j are compatible or not. The parameter λ is to adjust the relative ratio of the energy term and the penalty term. This energy function can be extended to include hairpin loops and pseudo knots (Fig. 3.12) as well.

Prediction of protein secondary structures

The RNA secondary structure prediction is basically a sequence alignment problem because the secondary structures are formed by complementary base pairs. In contrast, the secondary structures in proteins do not contain such conspicuous sequence features, making the problem more difficult to approach. Typical secondary structure elements in proteins are α-helices and β-sheets, both of which are characterized by regular backbone structures stabilized by hydrogen bonds. For example, the backbone dihedral angles (ϕ, ψ), which are defined in Fig. 3.13, take values of around $-60°$ and $-50°$ for right-handed α-helices.

Before many 3D structures became known an *ab initio* prediction of protein secondary structures had been attempted. This was based on the analysis of secondary structure formations by statistical mechanics. To provide a flavour of statistical mechanics, let us consider a very simplified problem. It is the helix–coil transition of homopolymers— synthetic polypeptide chains of single amino acids. Suppose that the chain consists of N residues. Each residue is assumed to take either the helical (h) state or the coil (c) state. The probability of observing a specific configuration of the chain is proportional to the product of the statistical weight, or the Boltzman factor (see *Simulated annealing* in Section 3.1), which is assigned to each residue as follows:

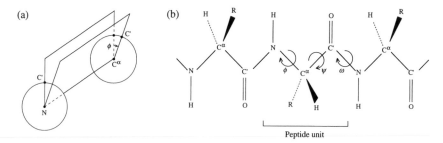

Fig. 3.13. (a) The definition of a dihedral angle and (b) the three backbone dihedral angles, ϕ, ψ, ω, in a protein. Because ω is around 180°, the backbone configuration can be specified by ϕ and ψ, for each peptide unit.

$$\begin{array}{cccccccccccccccc} \text{c} & \text{c} & \text{h} & \text{h} & \text{h} & \text{h} & \text{h} & \text{c} & \text{c} & \text{c} & \text{c} & \text{h} & \text{h} & \text{h} & \text{c} & \text{c} & \cdots \\ u & u & v & w & w & w & v & u & u & u & u & v & w & v & u & u & \cdots \end{array}$$

Thus, each c state has weight u, an h state that is adjacent to a c state has weight v, and an internal h state has weight w. Let us assume that the helix formation is cooperative, meaning that there is a tendency for the formation of separate clusters of helical and coil regions, which can be represented by the small value of v. This is the tendency of two different states or substances, such as oil and water, in a mixture to reduce their boundary areas. Since only the relative values are relevant, the weights are rewritten as: $u = 1$, $w = s$, and $v^2 = \sigma$. The weight for a coil region of any length becomes 1 and the weight for a helix region of length n becomes σs^{n-2}. For simplicity let us further assume that the polypeptide chain contains up to only one helical region and that the minimum length of a helix is three residues. Then, the partition function that enumerates all possible chain configurations is given by:

$$Z_N = 1 + \sigma \sum_{n=3}^{N} (N - n + 1)\, s^{n-2}$$

Once the partition function is known it is possible to compute various thermodynamic quantities, including the probability of any given residue of the chain being in the helical state (helix prediction).

The protein folding–unfolding transition is a cooperative phenomenon; the structural change occurs within a narrow range of the environmental variable, such as temperature. The helix–coil transition is less cooperative because of the one-dimensional nature of the homopolymers. The cooperativity of real protein transitions is likely to arise not from the local secondary structure formations, but from the long-range condensation of hydrophobic residues to form a core of the globule. Therefore, when statistical mechanics is to be applied to real proteins, it is necessary to consider long-range interactions rather than only neighbour interactions, as well as to consider 20 different kinds of amino acids rather than single amino acids. An *ab initio* prediction of protein secondary structures is no longer meaningful; it is equivalent to an *ab initio* prediction of protein 3D structures.

The Chou–Fasman method was the first empirical method for secondary structure prediction in proteins. It was based on statistical analysis of just 15 3D structures known at the time. They counted the observed frequencies of the 20 amino acids being in the helix, beta, and coil states, and defined the conformational parameters representing relative preferences to these states (see *Amino acid indices* in Section 2.3). Although the conformational parameters were based on simple single residue statistics, somewhat arbitrary rules were devised for nucleation and elongation of secondary structures. The first successful automated method was the GOR (Garnier–Osguthorpe–Robson) method, which was based on a more precise statistical analysis by incorporating correlations with up to eight residues on each side of the chain. In addition to these statistical methods, there were methods based on specific sequence patterns; notably, the 3.6 residue periodicity of hydrophobic and hydrophilic residues for amphipathic helices, which was used, for example, in Edmunson's helical wheel method. At present the prediction of protein secondary structures is achieved using more sophisticated methods in computer science.

Prediction of transmembrane segments

Membrane proteins form a major class of proteins that constitutes a quarter to a third of all the proteins encoded in the genome. They are the structural and functional components for the plasma membranes of cells and the special membranes that enclose organelles. Because membranes are hydrophobic in nature, the architecture of a membrane protein is sometimes termed as inside-out compared to a water-soluble globular protein which generally has a hydrophobic core surrounded by a hydrophilic shell. Figure 3.14 shows two examples of membrane proteins with a hydrophilic channel: bacteriorhodopsin in the plasma membrane, which consists of an α-helix bundle, and porin in the outer membrane of Gram-negative bacteria, which is a β-barrel protein. In comparison with globular proteins, the knowledge of 3D structures has been limited for membrane proteins because of the difficulties in experiments. However, the membrane-spanning α-helix appears to be a general structural motif for membrane proteins in the plasma membrane.

The pitch of an α-helix is about 0.15 nm per residue, while the apolar portion of the lipid bilayer is about 3 nm in thickness; thus about 20 hydrophobic residues are required for an α-helix to span the membrane. The cluster of hydrophobic residues can readily be identified as a candidate transmembrane segment by Kyte–Doolittle's hydropathy analysis. The amino acid sequence is converted into a numerical profile according to the hydropathy scale for the 20 amino acids (see *Amino acid indices* in Section 2.3) and the moving average is plotted for the segment of a given number of residues. The Klein–Kanehisa–DeLisi (KKD) method has automated this assignment by using discriminant analysis.

Discriminant analysis is a type of multivariate analysis that allows for the classification of sequence data with n attributes into m groups. Let us consider the simplest case of $n = 1$ and $m = 2$, in which a single attribute for the amino acid

(a) (b)

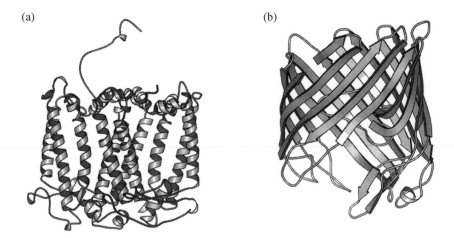

Fig. 3.14. Three-dimensional structures of membrane proteins: (a) α–helix bundle in the photosynthetic reaction centre (PDB:1PRC) and (b) β-barrel in an outer membrane protein, porin (PDB:1OMF).

segment of a given length is used to discriminate whether the segment is inside (group *I*) or outside (group *O*) of the membrane. Let *x* denote an attribute vector that characterizes the sequence data, and let $P(I|x)$ and $P(O|x)$ be, respectively, the conditional probabilities that the segment is inside and outside, given that it has attribute *x*. The discrimination is made by comparing the conditional probabilities; namely, the segment is inside the membrane if:

$$P(I|x) > P(O|x)$$

According to Bayes' theorem, the conditional probability $P(I|x)$ can be computed as:

$$P(I|x) = \frac{P(x|I)P(I)}{P(x|I)P(I) + P(x|O)P(O)}$$

and $P(O|x)$ can be computed in a similar way. Here $P(I)$ and $P(O)$ are the prior probabilities for the two groups, which may be estimated by looking at the relative abundance of each group in known proteins. Thus, the analysis involves the probability distribution of the attribute *x* in the two groups, $P(x|I)$ and $P(x|O)$. If the probability distribution is assumed to be normal, the discrimination rule becomes:

$$\frac{1}{\sigma_1}\exp\left\{-\frac{1}{2}\left(\frac{x-\mu_1}{\sigma_1}\right)^2\right\} > \frac{1}{\sigma_2}\exp\left\{-\frac{1}{2}\left(\frac{x-\mu_2}{\sigma_2}\right)^2\right\}$$

which can be represented by a quadratic function of x. Further, if the variances are assumed to be equal, that is $\sigma_1 = \sigma_2$, the result is a linear discriminant function. The KKD method utilizes the mean hydrophobicity in a segment of 17 residues as a single variable, and defines either the linear or the quadratic discriminant function from a training data set of known membrane proteins (see Fig. 2.12). This simple method still achieves over 90% of prediction accuracy.

Hierarchical neural network

In the 1950s the first artificial neural network (ANN) called the perceptron (Fig. 3.15(a)) was shown to be capable of discriminating patterns. The perceptron consists of two layers of nodes: multiple input nodes and a single output node. It receives an input vector (pattern) with the value of 1 or 0 at each input node and produces an output of 1 or 0 at the output node according to the following:

$$o = \begin{cases} 1 \text{ if } \sum_t w_i i_i > \theta \\ \\ 0 \text{ otherwise} \end{cases}$$

where i_i represents the input vector, o is the output, and w_i is the weight connecting input and output nodes. The output of 1 is produced when the weighted sum of inputs exceeds the threshold θ. A remarkable thing is that the perceptron can be trained by a process called supervised learning to produce outputs that best match target values. For example, the previous problem of discriminating transmembrane segments can be solved by the perceptron with 340 input nodes, representing each of the 20 amino acids at each position of the 17-residue segment. The training data set containing both true and false sequences is prepared and each sequence is repeatedly shown to the perceptron. The learning involves the adjustment of weights:

$$\Delta w_i = (t_p - o_p) i_{pi}$$

where t_p is the target value (given by the supervisor) and the suffix p indicates the particular pattern (sequence).

Because the discrimination is based on the linear combination of weights, the perceptron cannot solve problems that are not linearly separable. Thus, the perceptron was later extended to include additional layers, called hidden layers, between the input and output layers (Fig. 3.15(b)). The supervised learning for this hierarchical neural network with hidden layers is done by the back-propagation algorithm. First, the weights for the connections to the output layer nodes are adjusted to minimize the error between the target value and the actual output value. Using these weights, next, the weights for the connections to the previous hidden layer

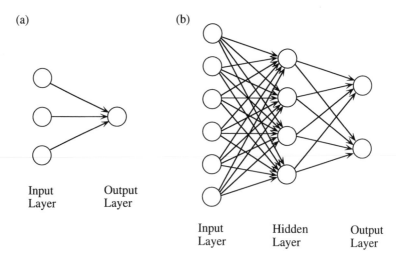

(a)

Input Layer   Output Layer

(b)

Input Layer   Hidden Layer   Output Layer

Fig. 3.15. Two architectures of the hierarchical neural network: (a) the perceptron and (b) the back-propagation neural network.

nodes are adjusted to minimize the error. This process is repeated up to the first hidden layer that connects to the input layer. Thus, the learning is performed by back-propagating the errors. The back-propagating neural network was successfully used in the prediction of protein secondary structures by Qian and Sejnowski and also in many other problems in computational biology.

Let us only consider the back-propagation learning algorithm for the simplest case, which is to optimize the weights for the output layer nodes. This involves minimization of the mean squared error defined by:

$$E = \frac{1}{2} \sum_j (t_j - o_j)^2$$

where t_j is the target value and o_j is the output value for the j-th node. The output value is determined from the weighted summation of the inputs i_i to this node with an activation function:

$$o_j = \frac{1}{1 + \exp(-\sum_i w_{ji} i_i + \theta_j)}$$

where w_{ji} is the weight for the connection between nodes i and j and θ_j is a threshold value. The activation function is a continuous function rather than the discrete function of 1 or 0 used in the perceptron. In the back-propagation algorithm the weights are optimized by the gradient descent, or by computing the derivative of the optimization function E:

$$\Delta w_{ji} = -k \frac{\partial E}{\partial w_{ji}}$$

where k is a constant. This derivative, which is readily computable, defines the learning rule for the output layer nodes.

Hidden Markov model

The finite state automaton that was used for the simultaneous search of multiple patterns (see *BLAST algorithm* in Section 3.1) is a deterministic automaton, where the state transition is uniquely determined by the input. In contrast, a hidden Markov model (HMM) is a probabilistic finite state automaton, where the state transition occurs according to the transition probability. HMM is a Markov chain because the probability of a state is dependent only on the directly preceding state without carrying the history of further preceding states. HMM, as well as ANN, has been successfully applied to many pattern recognition problems such as speech recognition.

Figure 3.16 illustrates a hidden Markov model as implemented in problems of biological sequence analysis by Haussler and colleagues. Suppose we wish to find a sequence motif that consists of five amino acid residues. The nodes m_0 and m_5 are the initial and final states, respectively, and the nodes m_1 to m_5 represent the states at five residue positions. The nodes in the middle and top rows represent, respectively, the states for insertions and deletions of residues. The insertion nodes i_0 to i_4 are placed before the five residue position nodes, and the deletion nodes d_1 to d_4 are placed to bypass the residue nodes. The arrows between nodes are associated with the transition probabilities, which are like the connection weights in ANN. Furthermore, the residue position nodes (m_1 to m_5) and the insertion nodes (i_0 to i_4) are associated with the output probabilities indicating which of the 20 amino acids are likely to occur. The output probabilities are hidden in the state transition diagram, hence, the name 'hidden' Markov model.

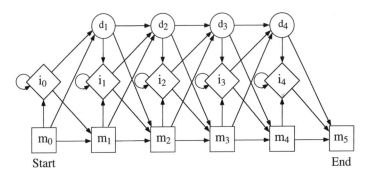

Fig. 3.16. A hidden Markov model for sequence analysis.

A hidden Markov model can be viewed as an extension of the concept of a profile (see Fig. 2.12). A profile is a numerical table containing, in this case, probabilities of observing different amino acids at each of the five residue positions. The profile thus corresponds to a collection of output probabilities at the residue position nodes (m_1 to m_5) in HMM. A profile may contain a deletion as the 21st amino acid type, but insertions cannot be included. HMM incorporates insertions and deletions more explicitly. Further, HMM is more general because it contains information regarding correlations between states; for example, the probability of observing two successive deletions. A profile can be constructed by the optimal alignment of a set of sequences, which is then used to count residue frequencies. In HMM a particular sequence is a series of states, so that a group of sequences can be used to optimize the transition probabilities and output probabilities and to construct the model that best characterizes the group. The length of the model, which is five in Fig. 3.16, may also be optimized.

This optimization is a search against the space of hidden Markov models for the model with the highest performance. The performance is measured in the following ways. Given a set of sequences $s(1)$, $s(2)$, ..., $s(n)$ in the training data set, how well a model fits the data set is represented by the joint probability of observing each sequence according to the model:

$$\text{Prob(dataset} \mid \text{model)} = \prod_{j=1}^{n} \text{Prob}(s(j) \mid \text{model})$$

This is called the likelihood of the model, and the principle of maximum likelihood can be used to measure performance. Alternatively, since Prob(data set | model) is the conditional probability of observing the data set, given a specific model, the Bayesian concept of maximizing the posterior probability can be used:

$$\text{Prob(model} \mid \text{dataset)} = \frac{\text{Prob(dataset} \mid \text{model)Prob(model)}}{\text{Prob (dataset)}}$$

where Prob(model) is the prior probability and Prob(dataset) is a normalization constant. This concept is related to the principles of minimum description length (MDL) and maximum parsimony (see *Phylogenetic analysis* in Section 3.1), which incorporate the complexity of a model as represented by the prior probability. It is better to find an approximate fit of the data with a simple model than to precisely fit the data with a far more complex model.

Different learning algorithms can actually find the best hidden Markov model. An iterative procedure, which is a type of expectation maximization (EM) algorithm, starts from an initial model. According to the current model, all possible trajectories for each sequence in the training data set are used to estimate new transition and output probabilities, which are then used to create a new current model. This iterative procedure will find a locally optimal solution. The expectation maximization (EM) algorithm can be combined with a stochastic procedure called Gibbs sampling, which is similar to Metropolis Monte Carlo simulation and simulated annealing (see *Simulated annealing* in Section 3.1), to better cope with an incomplete set of data.

Formal grammar

The nucleotide sequence or the amino acid sequence is a symbol string written with a limited number of letters, either four or twenty. The symbol string is supposed to contain biological meanings and in fact important words with special meanings are known as sequence motifs. In addition to the dictionary of sequence motifs, we would like to know the grammar, if any, in the language of biological sequences. Formal grammar as formulated by Noam Chomsky is defined by a vocabulary of terminal and non-terminal symbols and also by a set of rewrite rules. Terminal symbols cannot be rewritten, while non-terminal symbols, one of which is a start symbol, can be rewritten. Depending on the restrictions on the rewrite rule, Chomsky classified formal grammar hierarchically into four levels. Let us use upper-case letters such as A, B for non-terminal symbols, lower-case letters such as a, u, v for terminal symbols, and Greek letters such as α, β for any symbols.

Regular grammar is the simplest or lowest in the hierarchy with the most restrictive rewrite rule:

$$A \rightarrow aB$$

or

$$A \rightarrow a$$

Regular grammar is equivalent to regular expression in UNIX, and mathematically it can be represented by three operations: concatenation to connect symbol strings; disjunction to select from alternative symbol strings; and Kleene closure, which is conjunction to join symbol strings for any number of times including zero. The next level is context-free grammar, which is defined by the rewrite rule:

$$A \rightarrow \alpha$$

where there is no restriction on the right-hand side. By somewhat relaxing the restriction on the left-hand side as well, the rewrite rule:

$$uAv \rightarrow u\alpha v$$

defines context-dependent grammar. The most general grammar without any restrictions on both sides of the rewrite rule:

$$\alpha \rightarrow \beta$$

is unrestricted grammar. These rewrite rules may further be considered as stochastic processes to define stochastic grammars at different levels.

Languages generated by different levels of formal grammars are known to correspond to different classes of automata. A language in regular grammar is equivalent to a finite state automaton, a language in context-free grammar is equivalent to a pushdown automaton, and a language in context-dependent grammar is equivalent to a linear-bounded automaton. A language in unrestricted grammar is equivalent to Turing machine, which is said to be recursively innumerable. As we have seen, consensus sequence patterns are often represented by regular expressions (see *Protein families and sequence motifs* in Section 2.3). In fact, the sequence features such as motifs, profiles, and linear hidden Markov models are at the level of regular grammar. In order to incorporate RNA loop structures and palindromic DNA structures, context-free grammar is required. Furthermore, in order to represent RNA pseudo knots, context-dependent grammar is required. Beside these trivial examples, it is unclear whether formal grammars and equivalent methods are of practical use in understanding the hierarchy of, say, gene function expression and molecular structure formation.

Prediction of protein 3D structures

Knowledge based predictions of protein 3D structures can be classified into two categories: comparative modelling and fold recognition. When significant amino acid sequence similarity is found to a protein in the database of known 3D structures, it is better to use the knowledge of this particular protein structure. This is called comparative modelling in which a model structure is constructed by comparison with the reference structure. Differences in amino acid sequences may be adjusted for by minimization of conformational energy, as in the case of predicting 3D structure perturbations in protein engineering (see *Thermodynamic principle* above). In contrast, fold recognition is a method to predict the most compatible 3D fold among the known folds for a given amino acid sequence, even if there is no significant sequence similarity. This is based on the hierarchical nature of sequence similarity (superfamily) and structural similarity (fold); proteins without apparent sequence similarity may correspond to the same fold (see *Classification of protein 3D structures* in Section 2.3). Fold recognition requires the combined knowledge and analysis of all available protein structures.

Figure 3.17 illustrates the procedure of 3D–1D alignment, which is a simple way of recognizing folds, as presented by Eisenberg and coworkers. First, an amino acid residue in the 3D structure is classified into one of the 18 environmental classes, which are defined by the three main-chain states for the secondary structures and by the six side-chain states for the combination of buried/exposed and polar/apolar states. Thus, a 3D protein structure can be converted to a symbol string, each residue represented by a symbol for one of the environmental classes. Second, the preferences of different amino acids in different environmental classes are empirically determined from observed frequencies in a library of known protein structures, and they are represented by a matrix of 3D–1D scores. Third, using this score matrix, a protein structure (a symbol string) can further be converted to a 3D profile representing the preference of each amino acid at each residue position. Thus, a library of known structures has become a library of 3D profiles. A search against the standard profile library, which is a motif library, will identify matching functional sites for a given sequence. Similarly, a search against the 3D profile library will identify matching 3D structures for a given sequence. This process is thus called 3D–1D alignment.

Threading is a more sophisticated way of directly evaluating the effects of the three-dimensional environment. Conceptually, it is the threading of a given amino acid sequence through the hole made by the 3D topology of a polypeptide chain fold in the real protein structure. This process is also an alignment where the optimality is measured by an empirical potential energy function. The design of an appropriate potential function is most critical in the performance of threading methods. It generally includes pairwise potentials for side-chain interactions, scores for buried hydrophobic residues, and additional terms which are derived from the analysis of known protein structures. It may also include *ab initio* type potentials based on statistical mechanics. As the number of know structures continues to

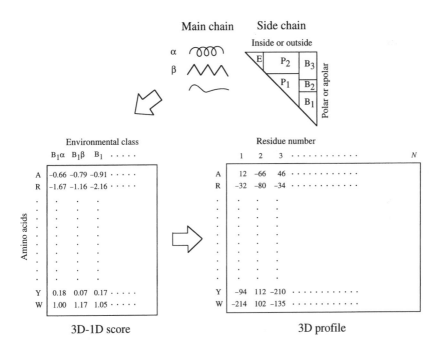

Fig. 3.17. The 3D–1D method for prediction of protein 3D structures involves the construction of a library of 3D profiles for the known protein structures.

increase (see Fig. 2.1), threading has become the state-of-the-art prediction method, incorporating both comparative modelling and *ab initio* prediction methods.

It has been argued that the number of folds in all the proteins that exist in nature is limited, maybe around one thousand. Once all of these folds are determined experimentally, so the argument continues, the protein folding problem is virtually solved (see *Grand-challenges in post-genome informatics* in Chapter 1) because of the success of fold recognition methods. However, this is a misleading argument when structure prediction is considered not as a goal of obtaining shape but as a means of understanding the protein's function. A fold represents a structural similarity only among a functionally divergent set of proteins. It remains to be seen whether fold recognition and associated methods will ever achieve the accuracy of distinguishing individual protein structures within the same fold. Furthermore, the precise prediction of a single molecular structure may not be sufficient for understanding function; improvements of molecular docking methods are also required.

Gene finding and functional predictions

Thus far, we have mostly considered the problems of structure predictions, which are better defined problems than functional predictions. Structure is a distinct entity and the accuracy of prediction can be measured objectively. In contrast, function is an empirical, often ambiguous concept, and it also depends on the degree of detail determined by subjective interests. The problem here in relation to sequence analysis involves the prediction of functional sites or functional units in single molecules. Table 3.2 shows typical examples of functional sites, which are the sites for interactions with other molecules. The most important functional units in a DNA sequence are transcription units and translation units to generate RNA and protein sequences. In the next chapter, the problems of functional predictions are treated at a higher level of abstraction, in terms of the function of a biological system as a whole. Functional predictions are all knowledge based; there is no *ab initio* functional prediction method. In general, functional predictions require accumulation and utilization of different pieces of knowledge, some of which may be just statistical features without much relevance to biological mechanisms involved. A case in point is the gene finding problem, which is to find genes, or predict coding regions, for functional proteins and RNAs in a genomic sequence.

The prediction of protein coding regions is a relatively easier task in prokaryotes because there is no phenomenon of splicing. An open reading frame (ORF)— a contiguous stretch of bases without a stop codon— can be considered to contain an entire protein coding region. First, all ORFs above a given threshold length are identified in the six reading frames, three on each strand. Then each ORF is examined to see if it corresponds to a gene based on judgement of combined pieces of evidence. The strongest evidence is significant sequence similarity to a well-characterized gene in the database. When no sequence similarity is found, an ORF

Table 3.2. Examples of functional sites

Molecule	Processing	Functional sites	Interacting molecules
DNA	Replication	Replication origin	Origin recognition complex
	Transcription	Promoter	RNA polymerase
		Enhancer	Transcription factor
		Operator and other prokaryotic regulators	Repressor, etc.
RNA	Post-transcriptional processing	Splice site	Spliceosome
	Translation	Translation initiation site	Ribosome
Protein	Post-translational processing	Cleavage site	Protease
		Phosphorylation and other modification sites	Protein kinase, etc.
		ATP binding sites	
	Protein sorting	Signal sequence, localization signals	Signal recognition particle
	Protein function	DNA binding sites	DNA
		Ligand binding sites	Ligands
		Catalytic sites	Many different molecules

can still be considered gene-like according to some statistical features, such as the three-base periodicity that arises from the bias in the third codon position and the tendency for higher G + C content in coding regions. Such evidence has to be augmented by additional evidence, notably signal sequence patterns for translation initiation and promoter binding sites in relation to the possible initiation codons.

In higher eukaryotes the gene finding becomes far more difficult because it is now necessary to combine multiple ORFs to obtain a spliced coding region. Alternative splicing is not uncommon, exons can be very short, and introns can be very long. Furthermore, sequence features are less conserved and more spread out, reflecting the complexity of regulatory mechanisms and the diversity of interacting molecules. Methods have been proposed to find genes in human genomic sequences using, for example, hierarchical neural networks and hidden Markov models. These methods are expected to improve as more knowledge is accumulated, especially from the comparison of genomic sequences and full-length cDNA (transcribed mRNA) sequences.

Like any other knowledge based predictions, the prediction of functional sites requires a well-verified data set of known sequences. Once the data set is prepared, the sequence features that characterize the data set can be extracted by various methods described in this chapter, such as multiple sequence alignment, discriminant analysis, neural networks, and hidden Markov models. In practice, however, it is not an easy task to manually collect and verify the data according to experimental evidence. An automatic approach is to couple the extraction of sequence motifs with the grouping of a sequence database (see *Protein families and sequence motifs* in Section 2.3). A group of sequences that are collected by global sequence similarity may be assumed to form a superfamily of functionally related proteins. If the group contains sufficiently divergent sequences, it may reveal locally conserved sequence features, which can then be used to extend the group to those sequences without apparent global similarity.

Expert system for protein sorting prediction

Table 3.2 shows that many problems of functional predictions are related to the processes of genetic information expression— transcription, post-transcriptional processing, translation, and post-translational processing. The final step of information expression for proteins involves their sorting to appropriate locations within the cell or to outside of the cell, as illustrated in Fig. 3.18. The prediction of protein localization sites requires combined knowledge of time-dependent and space-dependent processing by cellular machineries, together with the sorting signals of transported proteins. Because this type of knowledge is still fragmentary with frequent additions and modifications by new experimental results, the prediction of protein sorting does not fit well in the procedural description of conventional programming languages. It is expected to be much better handled by an expert system.

An expert system is an application of artificial intelligence, which is intended to aid problem solving in a narrow but realistic area by computerization of human expertise. Expert systems have been used in medical and biological applications since the late 1970s; early expert systems include DENDRAL for molecular structure determination in organic chemistry and MYCIN for medical diagnosis of

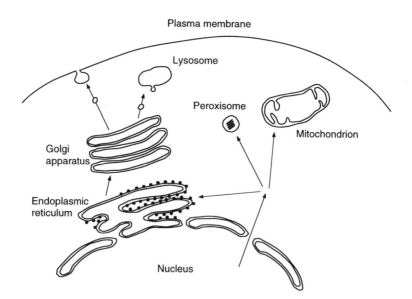

Fig. 3.18. A schematic illustration of protein sorting in animal cells.

bacterial infections. An expert system stores domain-specific knowledge and heuristics of human experts in a knowledge base, which is usually a collection of 'if–then' rules, or production rules. An expert system has the ability to infer new facts from stored knowledge and incoming facts, which is realized by an inference engine. The 'if' portion of a rule describes a set of conditions and the 'then' portion specifies a set of actions. Whenever some facts exist in a working memory, the inference engine matches these facts against the conditions of the rules and determines which actions are to be taken. Thus, from the user's point of view, the programming simply involves specification of rules and facts, which is very much like entering rules and facts in a deductive database (see *Deductive databases* in Section 2.2).

In Gram-negative bacteria, such as *Escherichia coli*, there are five cellular localization sites: cytoplasm, inner membrane, periplasm, outer membrane, and cell exterior (secreted). An expert system named PSORT was developed by Nakai and Kanehisa to evaluate different signals in the amino acid sequence and to combine the evaluations in a decision tree that mimics the actual sorting process. Most rules were based on experimentally identified sequence motifs, such as a signal sequence and its cleavage site, transmembrane segments, and a lipoprotein anchor site, all of which were refined by discriminant functions and profiles. In addition, some rules were based on statistical features, such as differences in amino acid compositions to distinguish between periplasmic and outer membrane proteins. In eukaryotic cells the sorting process is far more complex and involves a number of localization sites, but it was possible to take a similar approach.

In retrospect, there was a major difficulty in this expert system. Every rule in an expert system can be associated with a numerical quantity, called a certainty factor, that represents a relative preference among many rules. In principle, each certainty factor should be given *a priori* by an expert, but in practice it is an adjustable parameter which is like a connection weight of the artificial neural network. The addition or modification of a rule is not a trivial task because it requires manual readjustments of all the certainty factors to achieve optimal overall prediction accuracy. The current version of PSORT is no longer an expert system; it is a standard program written by a procedural language to better cope with this global optimization. Despite the promises made by artificial intelligence, an expert system in the form of a production system has not been very successful, at least in the biomedical applications. Conceptually it will be difficult for a machine to exceed human experts within a narrow domain of knowledge and with a limited amount of data. What a machine can do better is investigate, in a less detailed way, a large amount of data and knowledge that encompasses many different domains. With this prospect, we will consider a logic based approach in the next chapter.

4

Network analysis of molecular interactions

4.1 Network representation and computation

Level of abstraction

The central dogma of molecular biology summarizes the flow of genetic information expression at the sequence level:

$$DNA \rightarrow RNA \rightarrow Protein$$

In a traditional view, the thermodynamic principle and the structure–function relationship together establish an additional flow of genetic information expression for a single protein molecule:

$$Sequence \rightarrow Structure \rightarrow Function$$

The additional flow is spontaneous under proper physiological conditions; thus, the genome (DNA) contains virtually all necessary information for protein function.

Of course, this view is too simplistic and too reductionistic. Any biological function involves a network of interacting molecules. The information about molecular interactions is just as important as the information about single molecules. Therefore, it is necessary to understand the following information flow:

$$Interaction \rightarrow Network \rightarrow Function$$

which also implies the analysis of biological functions at a higher level of abstraction—the level of molecular networks rather than the level of single molecules (see Fig. 1.13). It is unlikely that the genome contains information about all necessary molecular interactions needed to make up life. The analysis inevitably involves space- and time-dependent behaviours of molecular interactions and reactions, both in terms of the physico-chemical principles and the biological constraints.

After the genome, it is natural to proceed to the analysis of the transcriptome and the proteome, which represent complete gene expression profiles at the mRNA and protein levels, respectively. The transcriptome and the proteome are likely to contain rich information on gene regulatory networks and protein interaction networks. Together with the complete sequence information in the genome, let us embark on the exploration of post-genome informatics.

Molecular networks

At the level of single molecules, the information regarding biological function is considered to be encoded in the sequence information, which is the linear arrangement of nucleotide or amino acid units. At the level of molecular networks, the information regarding biological function is considered to be encoded in the network information of interacting molecules. Here we define the term network in a general context. As shown in Fig. 4.1, a network is composed of both elements and binary relations. An element is either a molecule or a gene, and a binary relation is a molecular interaction, a genetic interaction, or any other relation between two elements. Conceptually, the building blocks and the connections, or the wiring, between them are both required to make up a biological system (see *Reductionism in biology* in Chapter 1). They are represented here by the elements and the binary relations, respectively. One may argue that the real world data can be better represented by considering ternary, quaternary, and even higher relations, but for the sake of logical simplicity and computational efficiency we impose the view of binary relations only. A higher relation is a composite entity, or a type of network, derived from a set of binary relations.

Figure 4.1 contains two broad categories of networks: one represents biological knowledge and the other is computed from binary relations. A pathway is a network of interacting molecules, representing the accumulated knowledge of various aspects of cellular processes, such as metabolic pathways, signal transduction pathways, cell cycle pathways, developmental pathways, and many other regulatory pathways. An assembly is another network of interacting molecules, in which molecules are

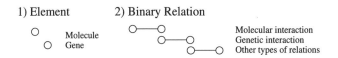

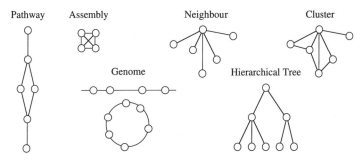

Fig. 4.1. Network representation. A network (graph) consists of a set of elements (vertices) and a set of binary relations (edges). Biological knowledge and computational results are represented by different types of network data.

known to form a stable complex, such as in molecular machinery or subcellular organization. The distinction between pathways and assemblies is sometimes blurred. For example, a multi-subunit enzyme complex, which is a molecular assembly, can be a node in a metabolic pathway, and the molecular complex formation can be better treated as part of a molecular reaction pathway. A genome is a network of genes, which represents the physical ordering of genes in the linear or circular chromosome. In comparison to the knowledge of a pathway or an assembly, which usually requires a series of intensive experiments, the knowledge of a genome is readily available by whole genome sequencing and computational gene finding.

The other types of networks shown in Fig. 4.1 are computational derivations of binary relations. A neighbour is a representation of similarity relations for a given molecule or a given gene. The result of a sequence similarity search is a typical example of a neighbour network where a query sequence is linked to multiple sequences by the binary relations of sequence similarity. We do not consider, in this case, the quantitative score of sequence similarity. Given a proper threshold score, the similarity relation becomes a logical relation, meaning that the relation is either present or absent. A cluster is a representation of similarity relations for an entire set of molecules or genes. Typically, a cluster is obtained by a cluster analysis (see *Multiple alignment* in Section 3.1) in which all pairwise similarity scores are used to somehow define similarity groups. For example, in single-linkage clustering, the group is a collection of members in which each member is linked to at least one other member of the group by the similarity relation, namely, the binary relation defined by a given threshold similarity score. In complete-linkage clustering, the group is a collection of members where each member is linked to every other member by the similarity relation. A hierarchical tree can then be a computational result of hierarchical clustering where the threshold scores for sequence similarity are progressively modified to define higher-level groups. A hierarchical tree can also be used to represent biological knowledge, such as the hierarchy of gene functions (see Table 2.4).

Graphs

At the level of single molecules, computerized sequence analysis has been extremely powerful for extracting biological information from sequence information. Ingenious methods have been developed for rapid sequence comparisons and for identifying characteristic sequence patterns as described in Chapter 3. At the level of molecular networks, computerized network analysis is also expected to play a major role in understanding biological information. Mathematically, a network is a graph, which is a collection of vertices (nodes) and edges that connect vertices. We have used the terms elements and binary relations, respectively, for vertices and edges. Therefore, the mathematical procedures involved are very much related to graph algorithms. Basically, there are two types of graph algorithms of interest: graph isomorphism and path computation.

A graph G with a set of vertices V and a set of edges E is denoted by:

$$G = (V, E)$$

In our problem of molecular networks, vertices are named—V is a finite series of vertices—so that G is a labelled graph. Note that all vertices need not be connected in a graph. In a connected graph there is always a path between any pair of vertices via one or more successive edges. A complete graph is fully connected; it has an edge between every pair of vertices. To incorporate various biological constraints of molecular networks, graphs may require different properties. A directed graph has one-way edges, and can represent an irreversible molecular reaction pathway. In a weighted graph, weights (costs) are assigned to the edges in order to, for example, distinguish between activation and inhibition in a signal transduction pathway. A weighted graph can also be used to define similarity scores or distances between vertices in the clustering procedure.

Let us consider a graph defined by the set of vertices A B C D E F and the set of edges between these vertices AB BC BD CE DE EF. In a standard representation, a graph can be drawn by marking points for the vertices and drawing lines for the edges. Figure 4.2(a) shows this representation, which is most intuitive and suitable for visualizing topological features of the graph. However, this is not the only representation of a graph. The same graph is represented by a linked list in Fig. 4.2(b), where each of the vertices is associated with information on connected vertices. Alternatively, the same graph is represented by a matrix of all pairs of vertices in Fig. 4.2(c), where 1 or 0 is entered to represent the presence or absence of an edge between vertices. It is interesting to note here that the concept of binary relations is naturally linked to the concept of a graph.

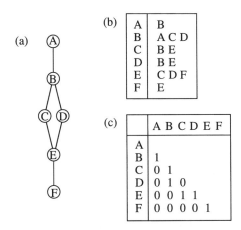

Fig. 4.2. Representation of the same graph by: (a) a drawing of nodes and edges, (b) a linked list, and (c) an adjacency matrix.

Two graphs:

$$G_1 = (V_1, E_1) \text{ and } G_2 = (V_2, E_2)$$

are said to be isomorphic when they have the same number of vertices, and there is a way to relabel the vertices from one graph to the other graph—V_1 is a permutation of V_2 and vice versa—so that the edges in both graphs become identical.

Common subgraph

Figure 4.3 illustrates biological examples of comparing two graphs of molecular networks: pathway–pathway, pathway–genome, genome–genome, and cluster–pathway comparisons. The edges of these graphs may be directed and/or weighted. When relabelling of vertices is defined according to the gene name in the genome and its gene product name in the biochemical pathway, or according to the sequence similarity of gene pairs or protein pairs, the biological comparison problem is reduced to finding common subgraphs, or isomorphic pairs of subgraphs, between two graphs. This is similar, in spirit, to finding locally similar subsequences (local similarity) of two sequences in the sequence analysis.

For two given graphs G_1 and G_2, subsets of the edges:

$$E_1' \subseteq E_1 \text{ and } E_2' \subseteq E_2$$

define subgraphs:

$$G_1' = (V_1, E_1') \text{ and } G_2' = (V_2, E_2')$$

A maximum common subgraph is the isomorphic pair G_1' and G_2' for which the common cardinality (set size) is the largest. This common subgraph represents identical matches without insertions and deletions, using sequence alignment terminology, although matches may be split into unconnected portions.

In order to incorporate insertions and deletions, subsets of vertices:

$$V_1' \subseteq V_1 \text{ and } V_2' \subseteq V_2$$

with the same cardinality are considered—the same number of vertices are selected from each graph. A maximum common induced subgraph is the isomorphic pair of the subgraph of G_1 induced by V_1' and the subgraph of G_2 induced by V_2' for which the common cardinality is the largest. This concept of maximum common induced subgraph is applied to a biological relevant example as shown in Fig. 4.4.

Suppose we wish to know if two biochemical pathways from different species

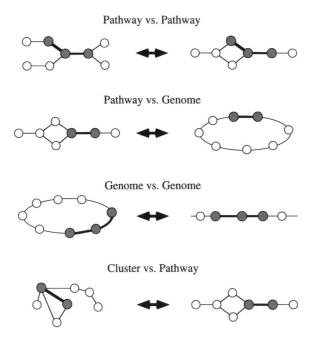

Fig. 4.3. Biological examples of network comparisons.

contain any similarity, based not simply on sequence similarity of individual molecules but rather on the similarity of wiring patterns among molecules. This might be called the problem of local pathway alignment. In Fig. 4.4(a), pathway 1 contains five molecules and pathway 2 contains six molecules, which are shown as the labelled vertices. The edges, excluding those in broken lines, represent activation steps (arrows) and an inhibitory step. If sequence similarities are found by comparison of all five by six sequences, they can be represented by additional edges as shown by thin broken lines. Then, the subpathways B → C and b → d are considered to be matched without any gaps. Furthermore, the subpathways B → C → D and b → d → → f can be considered to be similar by allowing a gap, which is represented by the additional edge d—f as shown with the thick broken line. Therefore, the local pathway alignment problem with gaps allowed can be approached by considering virtual edges between all non-adjacent vertices, namely, by considering two complete graphs with appropriate edge weights for gap penalties. Because these two complete graphs are then superimposed by relabelling (matching) vertices, the problem reduces to finding a specific feature in a single graph.

A feature is a set of mutually adjacent vertices, known as a clique. In Fig. 4.4(b), all possible relabelling in the two sets of vertices is represented by an induced graph with 5 × 6 vertices. The edges are present when the corresponding edges in the original graphs are compatible, namely, when they are two arrows or an arrow and a

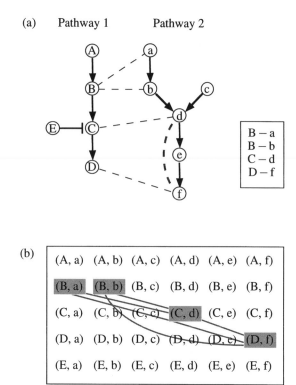

Fig. 4.4. Pathway alignment is a problem of graph isomorphism: (a) a maximum common induced subgraph and (b) a maximum clique.

virtual edge (gap). In this induced graph there are two cliques, but one of them formed by the mutually adjacent vertices (B, b) (C, d) (D, f) is optimal and this is what was identified as locally similar subpathways in Fig. 4.4(a). Therefore, the problem of finding a maximum clique can be transformed to and from the problem of finding a maximum common induced subgraph. This problem belongs to the category of NP-complete (nondeterministic polynomial time complete) problems, which are said to be solvable by a nondeterministic Turing machine in polynomial time, but which actually represent a class of extremely difficult problems with enormous computational complexity.

Heuristic network comparison

However hard a general problem is to solve in terms of computational complexity, real world problems can be practically solved by heuristic algorithms. In our problem of molecular network comparisons (Fig. 4.3), we wish to identify locally related regions, given correspondences of nodes (vertices) in two graphs. The number of actual correspondences is probably very limited; it is unrealistic to consider all possibilities of relabelling, namely, the product of the numbers of nodes in both graphs as we did in Fig. 4.4(b). For example, the number of correspondences in a genome–genome comparison is determined by how many sequence similarities are found, which is few. In the other types of real comparisons shown in Fig. 4.3 the number is even smaller. Thus, when two graphs are viewed as being linked by correspondences (additional edges) as shown in Fig. 4.4(a), our problem may be solved by seeking clusters of those correspondences. This is reminiscent of the local sequence alignment problem shown in Fig. 4.5(b), where a cluster of colons (:) corresponds to a locally similar region. Based on this idea, a practical algorithm for network comparison was developed by our group.

Let us consider two graphs:

$$G_1 = (V_1, E_1) \text{ and } G_2 = (V_2, E_2)$$

and a set of correspondences (Fig. 4.5(a)). In general, one node in one graph may correspond to multiple nodes in the other—the correspondences can be many-to-many—but here all correspondences are represented by pairwise (binary) relations. If the set contains n correspondences, the problem is to cluster these n data points according to a certain measure of distance. Since each data point represents correspondence between a node in G_1 and a node in G_2, the distance of two data points i and j may be defined by the following two distances:

$d_1(i, j)$ for the shortest path between nodes v_{1i} and v_{1j} in graph G_1, and

$d_2(i, j)$ for the shortest path between nodes v_{2i} and v_{2j} in graph G_2

First, each correspondence is considered as an individual cluster; thus, there are n initial clusters. Then the single linkage clustering is performed according to the following criterion of whether to merge two clusters C_i and C_j:

$$\delta(i,j) = \begin{cases} 1 \text{ if } \min_{r,s}\{d_1(r,s) \mid r \in C_i, s \in C_j\} \leq 1 + Gap_1 \text{ and} \\ \quad\quad \min_{r',s'}\{d_2(r',s') \mid r' \in C_i, s' \in C_j\} \leq 1 + Gap_2 \\ 0 \text{ otherwise} \end{cases}$$

where Gap_1 and Gap_2 are gap parameters of non-negative integers. Thus, clusters are merged ($\delta = 1$) allowing a certain degree of gaps and mismatches.

Note that this graph comparison algorithm does not detect isomorphic subgraphs; rather it detects correlated clusters of nodes in two graphs, which are likely to be biologically more significant. The algorithm requires computation of the shortest path between two given nodes in a graph, which is described below.

(a)

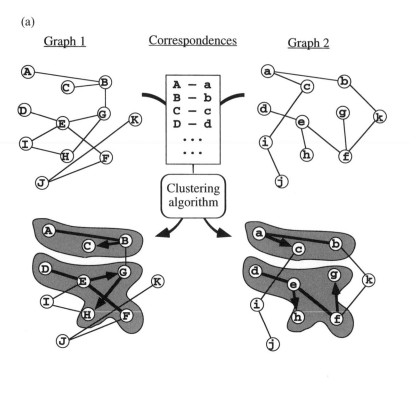

Fig. 4.5. A heuristic algorithm for biological graph comparison. It searches for clusters of correspondences, as shown in (a), which is similar in spirit to sequence alignment, shown in (b).

(b)

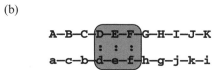

Path computation

A node is connected to another node when there is a path—either a direct path connecting the nodes by a single edge or an indirect path connecting one or more successive nodes by multiple edges. This definition is valid for both undirected and directed graphs. When node *i* is connected to node *j*, and node *j* is connected node *k*, then node *i* is also connected to node *k*. Thus, an edge *i* to *k* can be added if it does not exist. The graph that results from this procedure of adding all direct edges to each connected portion of a graph is called the transitive closure. The simple logic is implemented in Warshall's algorithm shown below:

```
for (j = 1; j <= N; j++)
    for (i = 1; i <= N; i++)
        if (a[i][j])
            for (k = 1; k <= N; k++)
                if (a[j][k]) a[i][k] = 1;
```

which is applied to an adjacency matrix $a(i, j)$, and N is the number of nodes. An element of the adjacency matrix is either 1 or 0, depending on whether two nodes are adjacent (directly connected) or not, as shown in Fig. 4.2(c). For simplicity the matrix is made symmetric and the diagonal elements are assigned 1. Obviously, this is an $O(N^3)$ algorithm. Precomputation of the transitive closure is useful to efficiently find, for example, if there is a path from one node to another.

The transitive closure does not, however, answer the question of which is the shortest path among many possibilities. The shortest path represents the smallest number of edges between two nodes in an unweighted graph, or it is based on the smallest sum of edge weights in a weighted graph. Actually, we have already encountered the shortest path problem in Fig. 3.2 for the sequence alignment problem, which was to find the shortest (optimal) path from the node in the upper-left corner to the node in the lower-right corner of the path matrix. The problem was solved by the breadth–first search in combination with a dynamic programming procedure in which only the shortest paths up to the current level are retained for computing the paths further down the tree. The same strategy, called Dijkstra's algorithm, can be applied to finding the shortest paths from a given node to each of the other nodes in both unweighted and weighted graphs.

To find the shortest paths for all the pairs of nodes in a graph, Dijkstra's algorithm may be applied V times starting at each node. However, an extension of the transitive closure, called Floyd's algorithm, is more efficient, which is still $O(N^3)$:

```
for (j = 1; j <= N; j++)
    for (i = 1; i <= N; i++)
        if (a[i][j])
            for (k = 1; k <= N; k++)
                if (a[j][k] > 0)
                    if (!a[i][j] || (a[i][j]+a[j][k] < a[i][k]))
                        a[i][k] = a[i][j]+a[j][k];
```

Here again the matrix element is 0 when the edge is absent, 1 when the edge is present in an unweighted graph, and appropriate positive values when the edge is present in a weighted graph.

The shortest path problem was to find an optimal set of edges, in terms of the sum of edge weights, that connects two nodes of a graph. Another optimization problem is to find an optimal collection of edges, again in terms of the sum of edge weights, connecting all the nodes of a graph. The resulting graph is called the minimum spanning tree, where a tree is a special graph with no cycles. We do not discuss further the algorithms for solving this problem, but only mention here the relationship of the minimum spanning tree of a complete graph to the single-linkage hierarchical cluster analysis. The cluster analysis involves a set of N points in a geometric space where the distance is defined for all the pairs, which is a complete weighted graph with N nodes and $N(N-1)/2$ edges. The minimum spanning tree is a convenient way to represent the result of single-linkage clustering.

Binary relation and deduction

The basic strategy for computing the transitive closure was: 'when node i is connected to node j, and node j is connected node k, then node i is also connected to node k.' This is actually syllogism in logical reasoning. An edge (connection) is a binary relation, a path is a deduction step, and a graph is a whole network of possible deduction steps. Once different types of data and knowledge are organized in terms of binary relations, they can be used for automatic path computations; thus, there is potential for computerizing human reasoning steps. In fact, the concept of binary relations can be far more general; it is not limited to just molecular interactions and relations.

In Table 4.1, binary relations of relevance are categorized into three types: factual relations, similarity relations, and functional relations. The categorization of factual links is somewhat operational. Factual relations represent mostly trivial links between different database entries, which are stored as cross-reference information in the molecular biology databases. Similarity relations are computationally derived from comparison against the sequence or 3D structural database. The manipulation of such binary relations has already been discussed in the context of the database technology, especially, the deductive database technology (see *Deductive databases* and *Link-based integration* in Section 2.2). For example, searching a set of parent–child binary relations for ancestors is equivalent to searching a family tree. In the DBGET/LinkDB system indirect cross-references can be found by combining mul-

Table 4.1. Examples of binary relations

Type of relation	Contents	Examples
Factual relation	Links between database entries	Factual data and its publication information Nucleotide sequence and translated amino acid sequence Protein sequence and 3D structure
Similarity relation	Computed similarity Computed complementarity	Sequence similarity; 3D structural similarity 3D structural complementarity
Functional relation	Molecular reactions Molecular interactions Genetic interactions Chromosomal relations Evolutionary relations	Substrate–product relations Molecular pathways; molecular assemblies Positively co-expressed genes Negatively co-expressed genes Correlation of gene locations (operons) Orthologous and paralogous genes

tiple links, which is accomplished efficiently by a precomputed partial transitive closure as well as by a dynamic linking capability.

Figure 4.6 shows an example of path computation for a specific class of binary relations, namely, substrate–product relations in enzymatic reactions. For simplicity, a reaction involving multiple substrates and/or multiple products is decomposed into all possible substrate–product pairs. Thus, for a set of enzymes, a possible reaction space is represented by a graph in which nodes are chemical compounds and edges are labelled by enzyme names, namely, EC numbers plus additional identifiers to distinguish pairs in multi-compound reactions. Given two compounds, pyruvate (C00022) and L-alanine (C00041), the shortest path is found to be a single-step reaction by EC 1.4.1.1, but there are many other alternative paths involving two, three, and more enzymes as shown in the figure. Let us formulate this path computation in terms of logic programming (deductive database).

Suppose that enzyme E catalyses the chemical reaction converting compound X to compound Y, which is a fact represented by:

$$reaction\ (E, X, Y)$$

The rule for path computation can then be defined recursively as follows:

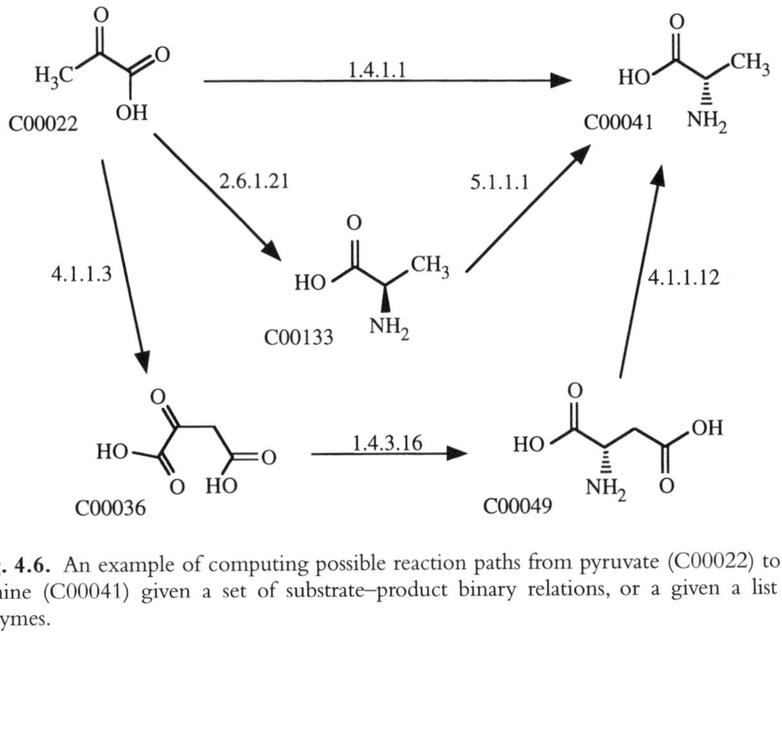

Fig. 4.6. An example of computing possible reaction paths from pyruvate (C00022) to L-alanine (C00041) given a set of substrate–product binary relations, or a given a list of enzymes.

$$path\ (X,\ Y,\ [E])\ \leftarrow\ reaction\ (E,\ X,\ Y)$$

$$path\ (X,\ Y,\ [E\ |\ EL])\ \leftarrow\ reaction\ (E,\ Z,\ Y),\ path\ (X,\ Z,\ EL)$$

where EL is a list of enzymes. There is a path from X to Y if there is a reaction from X to Y or if a reaction from Z to Y extends an existing path from X to Z. From a biological point of view, it often helps to include a hierarchical grouping of enzymes in this path computation, such as to assume wider substrate specificity by ignoring the difference of the last digit in the EC number hierarchy. The following rule:

$$reaction(E',\ X,\ Y)\ \leftarrow\ reaction(E,\ X,\ Y),\ group(G,\ E,\ E')$$

states that when enzymes E and E' are similar, belonging to the same group G, the reaction catalysed by enzyme E may also be catalysed by enzyme E'. This procedure, known as query relaxation, is useful because the EC number assignment by sequence similarity may not be accurate enough to determine substrate specificity, in which case it is necessary to consider all reactions in the same category. Query relaxation effectively expands the number of binary relations (edges) and the possible reaction space (graph) to be searched.

Another type of query relaxation is shown in Fig. 4.7, in which the hierarchical grouping of protein molecules (nodes) is used to deduce a network of protein–protein interactions. With the five binary relations shown there is no path from B to F. However, if proteins E and E' are considered to be functionally equivalent (group G), for example, according to the sequence similarity or the sequence motif, the path may be said to be virtually present. This reasoning step corresponds to introducing an additional edge to connect nodes in the existing graph. It is expected that human reasoning can be computerized by a method similar to path computation, or deduction, from a set of binary relations.

Implementation

In contrast to the factual and similarity relations, functional relations shown in Table 4.1 are largely uncomputerized in the traditional databases, except for the substrate–product relations of enzymatic reactions. We have presented a conceptual

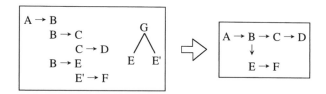

Fig. 4.7. Query relaxation. Nodes E and E' are considered to be equivalent according to the grouping G.

framework for computer representation and utilization of functional information. A diverse range of functional data should be represented in the form of binary relations in order to combine both computationally and logically among themselves and with other different types of binary relations. Let us now examine practical implementation of this concept in the KEGG (Kyoto Encyclopedia of Genes and Genomes) system.

The level of abstraction that we consider is the molecular network level. The data representation in KEGG is based on the view shown in Fig. 4.1. The elements are molecules and genes. The binary relations are molecular interactions, genetic interactions, and other relations between molecules or genes. The networks are the most unique features in KEGG, and include all the functional relations shown in Table 4.1. The actual implementation of different types of networks is summarized in Table 4.2. The pathway map contains information on metabolic pathways, regulatory pathways, and molecular assemblies. The genome map is a one-dimensional network of genes in the chromosome. The expression map represents environment-dependent and time-dependent expression of all the genes in the genome, which contains information on gene regulatory networks such as clusters of positively and negatively co-expressed genes. The orthologue group table (see Fig. 2.15) represents a set of orthologous genes for a functional unit in different organisms, such as for a molecular assembly or a pathway motif that is a conserved portion of a biochemical pathway. In addition, the table contains the positional correlation of genes in the chromosome, known as an operon structure, which often represents a set of co-regulated genes. Figure 4.8 shows a typical example where a set of positionally correlated genes in the genome corresponds to a functional unit in the metabolic pathway. This can be found in KEGG by a type of network comparison, a genome–pathway comparison.

Table 4.2. Network data representation in KEGG

Network type	KEGG data	Content	Representation
Pathway Assembly	Pathway map	Metabolic pathway, regulatory pathway, and molecular assembly	GIF image map
Genome	Genome map Comparative genome map	Chromosomal location of genes	Java applet
Cluster	Expression map	Differential gene expression profile by microarrays	Java applet
Neighbour Pathway Assembly Genome	Orthologue group table	Functional unit of genes in a pathway or assembly, together with orthologous relation of genes and chromosomal relation of genes	HTML table
Hierarchical tree	Gene catalogue Molecular catalogue Taxonomy Disease catalogue	Hierarchical classification of genes Hierarchical classification of molecules Hierarchical classification of organisms Hierarchical classification diseases	Hierarchical text

(a) *E. coli* genome

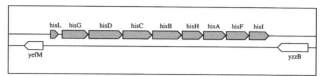

(b) Metabolic pathway

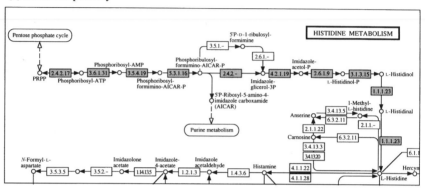

Fig. 4.8. Genome–pathway comparison, which reveals the correlation of physical coupling of genes in the genome (operon structure) and functional coupling of gene products in the pathway.

   The hierarchical tree network in Table 4.2 is a convenient way of summarizing knowledge of similarity relations, for example, the classification of protein molecules according to the sequence similarity (superfamily), the 3D structural similarity (fold), and the local sequence pattern (sequence motif). The hierarchical tree representation may be too restrictive for the complexity of biological relations to be fully appreciated, but it can provide an approximate view that captures the most important features of biological relations. The gene catalogue in KEGG represents functional hierarchies of genes according to the classification of biochemical pathways and assemblies. The molecular catalogues include functional and/or structural classifications of protein molecules, RNA molecules, and chemical compounds. Figure 4.9 shows the result of searching for clusters of a specific class of proteins, β/α barrel proteins, in the metabolic pathway, which is another example of network comparison in KEGG—a comparison between the hierarchical tree and the pathway. This reveals the existence of successive reaction steps catalysed by structurally similar enzymes, which suggests possible roles of gene duplications in the pathway formation.

   KEGG provides various computational tools to search and compare different types of networks (see Fig. 4.3) stored in the database. A more challenging computational problem is network prediction, which involves predicting the entire

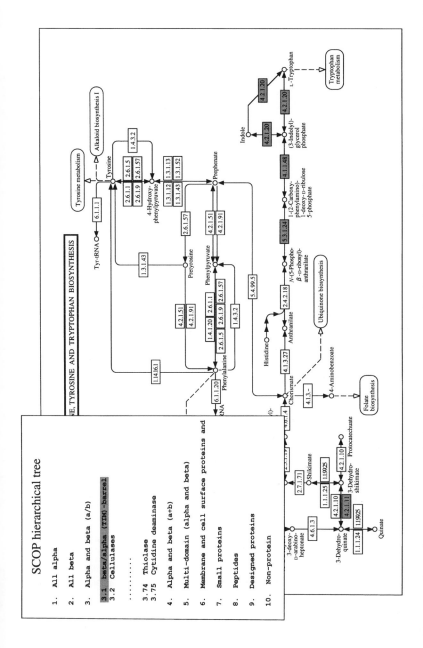

Fig. 4.9. Hierarchy–pathway comparison, which reveals the correlation of evolutionary coupling of genes (similar sequences or similar folds due to gene duplications) and functional coupling of gene products in the pathway.

Table 4.3. Grand challenge problems

	Protein folding problem	Organism reconstruction problem
Prediction	Structure prediction—to predict protein 3D structure from amino acid sequence	Network prediction—to predict entire biochemical network from complete genome sequence
Knowledge	Known protein 3D structures	Known biochemical pathways and assemblies
Knowledge based prediction	Threading	Network reconstruction
Ab initio prediction	Energy minimization	Path computation
Prediction of perturbed states	Protein engineering	Pathway engineering

biochemical network from the complete genome sequence. Table 4.3 compares the problems of structure prediction and network prediction, which are actually the two grand challenge problems in computational molecular biology termed, respectively, the protein folding problem and the organism reconstruction problem (see *Grand challenges in post-genome informatics* in Chapter 1). Although in principle the protein structure is computable from the amino acid sequence information alone, in practice it is not feasible. Similarly, although in principle the entire biochemical network may be computable from the genomic information alone, in practice it is impossible. However, with the accumulation of known 3D structures, the knowledge based prediction of protein structures has become more successful and practical. Thus, the knowledge based prediction of biochemical networks is also expected to become of practical use with the accumulation of known biochemical pathways and assemblies. KEGG aims at providing a good reference database of all known biochemical networks, which is to be utilized, as in the case of the reference database of protein 3D structures, for knowledge based predictions.

4.2   Principles of biochemical networks

Metabolic network

Life arose on Earth four billion years ago bringing to an end to the prebiotic era which had begun one billion years earlier with the formation of the solar system. The initial membrane-bound, self-replicating forms of life acquired energy from oxidation of inorganic materials (chemolithotrophy) that had been formed in the prebiotic era, but eventually oxygen–producing photosynthesis evolved to acquire energy from sunlight (phototrophy). Together with these developments different forms of life evolved able to acquire energy from oxidation of organic compounds (chemoorganotrophy), especially from respiration which uses oxygen as an oxidizing agent. This is the likely scenario of how life evolved. In fact, life on Earth as a whole should be viewed as a huge biochemical network that has evolved and will continue to evolve with Earth, with the solar system, and ultimately with the expanding universe.

At the level of individual organisms such a huge biochemical network is considered an environment. The cellular forms of organisms acquire energy, carbon, and other necessary materials from the environment and utilize them to maintain their own biochemical networks. Metabolism is the most basic network of biochemical reactions, which generate energy for driving various cell processes, and degrade and synthesize many different molecules. There is a rough division of metabolism into intermediary metabolism and secondary metabolism: intermediary metabolism is a core portion of reaction pathways that are conserved among organisms; secondary metabolism is a more divergent set of reactions that are directly influenced by the environment. For example, the presence of foreign compounds or other environmental changes can cause microbes to generate new biodegradation pathways by mutations and gene transfers of enzymes. The generated pathways are at the level of secondary metabolism, whereby partially degraded intermediates are funnelled into the common pathways of intermediary metabolism. This architecture of metabolic pathways is somewhat reminiscent of the architecture of globular proteins, which consists of a conserved hydrophobic core to maintain the globule and a divergent surface area for interactions with other molecules.

Figure 4.10 shows the central portion of intermediary metabolism: glycolysis, the TCA cycle, and the pentose phosphate pathway. This network diagram can be viewed, approximately, as a directed graph where each node is a chemical compound with its name shown beside it and an edge is an enzyme-catalysed (not shown) chemical reaction between compounds. This diagram also contains biological details that will make mathematical analysis somewhat complicated. First, a reaction is not a binary relation of compounds; it can involve multiple substrates and multiple products. Second, all of the compounds involved in the reaction are not necessarily shown. In fact this type of network representation is used to illustrate only the conversion steps of major compounds with some additional features shown by small branches. Since here we wish to emphasize how energy is produced in the central pathways of intermediary metabolism, both in terms of the so-called high-energy bond in ATP and the reducing power of NADH and NADPH, such energy producing steps are included as small branches. Further, since we wish to illustrate how the six-carbon compound of glucose is converted to other important intermediary compounds, each node is associated with the number of carbons and again small branches for CO_2 and CoA are included to account for the equality of the number of carbons in the substrates and products. Third, because an edge represents an enzyme, two nodes can be connected either by a single edge for an enzyme catalysing reactions in both directions or by two separate edges when there are two different enzymes for respective directions, which is the case between D-fructose 6-phosphate and D-fructose 1,6-bisphosphate during glycolysis or gluconeogenesis.

Furthermore, these central pathways are never an isolated network. There are numerous connections with other metabolic pathways and also some connections

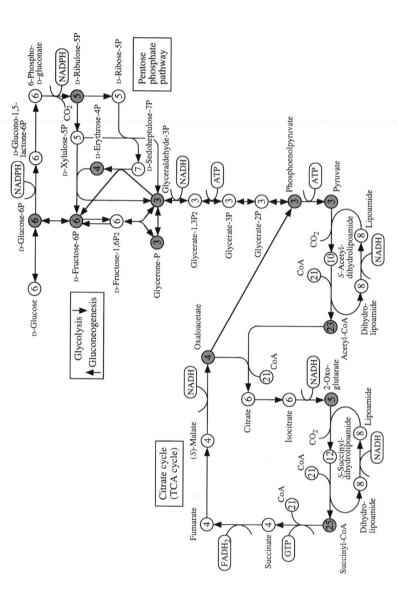

Fig. 4.10. Glycolysis, the TCA cycle, and the pentose phosphate pathway, viewed as a network of chemical compounds. Each circle is a chemical compound with the number of carbons shown inside.

with cellular transport systems. The shaded nodes in Fig. 4.10 are, what Neidhardt called, 12 precursor metabolites, which serve as the starting point for many important biosynthetic pathways in intermediary metabolism. It appears that one of the most important principles of the metabolic network is to rely on a selected number of intermediary compounds. There are hubs in the network where many routes converge in and diverge out. When details are abstracted, metabolism may be viewed as a network of trunk pathways between hubs. Thus, metabolic logic of using a selected number of intermediary compounds is related to a hierarchical organization of the metabolic network.

In the KEGG system the metabolic network of chemical compounds is implemented as a graph—a set of binary relations between compounds. In order to cope with the biological details mentioned above, a reaction containing multiple substrates and/or multiple products is split into all possible substrate–product pairs. In the current implementation, all the reactions are assumed to be reversible so that the graph is undirected. It is an option to restrict the set of binary relations, whether to consider all the compounds involved or only those main compounds that appear on the metabolic pathway diagrams in KEGG. Non-enzymatic reactions in the known pathways are also included in the set of binary relations, which is being expanded to further incorporate knowledge of organic chemistry. The path computation with this graph is implemented as a tool in KEGG, which can be used to search for possible reaction paths given either a starting compound or both a starting and an ending compound (see Fig. 4.6).

Genomic perspective

While metabolism is a network of chemical compounds, it is also a network of enzymes. This view is shown in another graph representation, Fig. 4.11, which depicts in detail only glycolysis among the central pathways of intermediary metabolism shown in Fig. 4.10. Here a node is an enzyme represented by a box labelled with its EC number. An edge is the connection of two enzymes which contains the compound in between—both a product of one enzyme and a substrate of the next enzyme. Because a network of enzymes is equivalent to a network of genes coding for the enzymes in each organism, this graph representation is most useful in superimposing the genomic information on the knowledge of metabolic pathways, which helps to deduce metabolism for each organism. At the time of writing this, complete genome sequences are publicly available for only about 20 organisms. Knowledge based prediction of biochemical networks is at a very early stage of development, but what follows are some insights gained thus far toward understanding the general principles that underlie both the genetic logic and chemical logic of the metabolic network.

The knowledge based prediction of metabolic pathways from the complete genome sequence involves the matching of genes in the genome against enzymes in the KEGG reference pathway diagrams. As shown in Fig. 4.11, an enzyme is

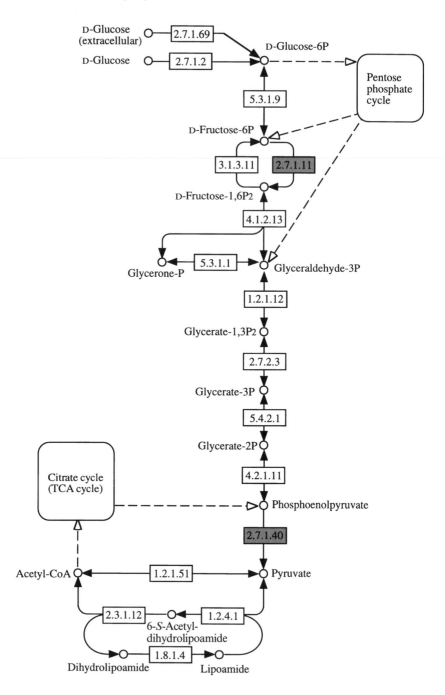

identified by the EC number in the pathway. Thus, it is first necessary to predict enzyme genes in the genome and assign EC numbers by a sequence similarity search against a well-annotated sequence database, such as the GENES database in KEGG. Once this is done, matching against the reference pathway is automatic and the result is conveniently shown in KEGG by colouring the enzymes (boxes) whose genes are identified in the genome. The connectivity of coloured boxes can then be used to judge if a specific pathway is present or absent, as well as if any enzyme is missing in an otherwise complete pathway. In the latter case, it is necessary to re-examine the gene assignment process and also to search for alternative paths (enzymes) by path computation of chemical compounds.

The matching of genes and pathways is actually a graph comparison—a genome–pathway comparison (see Fig. 4.3). Therefore, it additionally reveals the clusters of genes in the genome that are also clustered in the pathway (see Fig. 4.8). Such operon structures are prevalent in bacterial and archaeal genomes, but only limited, different sets of genes are in the operons in any given organism. Thus, combined knowledge of physically correlated sets of genes should be organized to understand functional correlations of genes, especially in metabolic pathways and molecular assemblies. This can be accomplished by another type of graph comparison, or genome–genome comparisons, for all completely sequenced genomes. Because a set of positionally correlated genes is likely to be co-expressed, the same set of orthologous genes in another genome is also likely to be co-expressed even if the genes are separately located. Furthermore, such co-expression analysis can be performed more directly and systematically by microarray and similar experiments for gene expression profiles. In this case, the graph comparison is a cluster–pathway comparison, where a cluster is a network of positively and negatively correlated genes under different conditions. The integrated analysis of these three types of graph comparisons—genome–pathway, genome–genome, and cluster–pathway—is expected to expand our knowledge by filling gaps of unidentified metabolic reactions and, more importantly, by clarifying regulatory mechanisms of metabolic pathways.

In glycolysis, shown in Fig. 4.11, there are three pieces of the pathway that are identified as positionally correlated sets of genes. The last reaction step from pyruvate to acetyl-CoA is catalysed by the products of the three genes for EC 1.2.4.1, 2.3.1.12, and 1.8.1.4, which are commonly found in an operon structure. In fact, these gene products form the molecular assembly of pyruvate dehydrogenase, a multi-subunit enzyme complex whose overall reaction is represented by EC 1.2.1.51. In some organisms, the genes for phosphofructokinase (EC 2.7.1.11) and pyruvate kinase (EC 2.7.1.40), which are shaded in the figure, are positionally

Fig. 4.11. Glycolysis viewed as a network of enzymes (gene products). Each box is an enzyme with its EC number inside.

correlated, suggesting that they are co-regulated. The distance between these enzymes' positions in the pathway may at first suggest that they are not functionally correlated. Interestingly, however, these enzymes catalyse the only two irreversible reaction steps in glycolysis, which are key steps in controlling the overall direction of the reaction: glycolysis or gluconeogenesis. It is also known that the product of phosphofructokinase, fructose 1,6-bisphosphate, stimulates the activity of pyruvate kinase by feedforward allosteric regulation. Thus, the overall regulation of glycolysis and gluconeogenesis is performed at both the gene expression level and the protein interaction level.

The six enzymes in between these two irreversible reaction steps tend to be clustered in the genomes of many organisms, although in most cases only subsets of the six are clustered. Furthermore, these six genes appear to form one of the most tightly coupled clusters of co-expressed genes, second only to the gene cluster of ribosomal proteins. This is seen in the analysis of microarray gene expression profiles of *Saccharomyces cerevisiae*, despite the fact that the genes are dispersed in the genome. Archaea do not generally have a complete glycolysis pathway, but most of the genes in this cluster are present although not necessarily in an operon. It appears that this section of glycolysis is the most conserved portion in intermediary metabolism, and might play roles in energy metabolism and pyruvate metabolism in chemolithotrophic and phototrophic organisms as well. Note, however, that an obligate parasite may lack the entire pathway of glycolysis including this conserved portion. The TCA cycle is completely or partially missing in many microorganisms, and again the piecewise organization is apparent. The TCA cycle as we know it now appears to be a highly elaborate system for energy metabolism that results from the assembly of different pieces of reaction pathways.

Protein–protein interaction network

Metabolism represents a relatively well-known part of the biochemical pathways, except for secondary metabolism, which can be explained by simple logic of chemical reactions among small compounds. In contrast, there are other various regulatory pathways that still need to be identified, especially from the analysis of complete genome sequences. These pathways involve protein–protein interactions, which are far more complicated than simple chemical reactions. Of course, the distinction between metabolism and regulation is not necessarily clear-cut. Macromolecular metabolism of messenger RNA synthesis (transcription) and protein synthesis (translation), for example, represents highly regulated pathways of macromolecular interactions. The signal transduction pathway can contain enzymatic reactions, such as phosphorylation, as well as small chemical substances, such as cyclic AMP and calcium ion. Therefore, it is necessary to consider all the different types of molecular interactions in order to fully understand the entire biochemical network.

Looking at all the different molecular interactions is an extremely complex task.

An abstraction that we take here is to consider just protein–protein interactions, or binary relations of proteins, which is probably most practical for linking genomic information and network information as well as integrating metabolism and regulation. Figure 4.12 explains this concept. There are direct protein–protein interactions such as binding interactions, including formation of macromolecular assemblies, covalent modifications of phosphorylation, glycosylation, and others, and proteolytic processing of polypeptide chains. As we have seen in Fig. 4.11, the metabolic network of chemical compounds can also be viewed as a network of enzyme molecules. In this case two proteins (enzymes) are considered to interact 'indirectly' via successive chemical reactions. Another important class of indirect protein–protein interactions is gene expression, where the message of one protein is transmitted to another protein via the process of protein synthesis from the molecular template (gene). Functional RNA molecules, such as tRNAs, rRNAs, and snRNAs, should also be included as variants of proteins. Thus, the protein–protein interaction network considered here is actually the interaction network of all the gene products.

Figure 4.13 illustrates our strategy to predict or reconstruct this generalized protein–protein interaction network from the complete genome sequence. First, the current knowledge of all biochemical networks in all organisms has to be stored in a reference database such as KEGG. Second, the knowledge based prediction is performed, where for each of the genes in the genome a matching component (gene

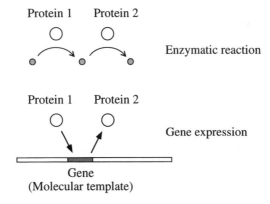

Fig. 4.12. A generalized concept of protein–protein interaction.

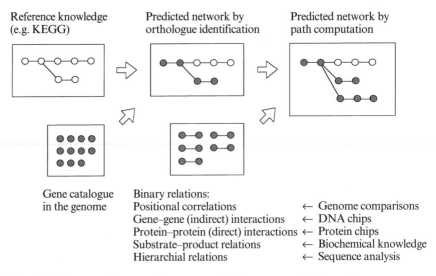

Reference knowledge
(e.g. KEGG)

Predicted network by
orthologue identification

Predicted network by
path computation

Gene catalogue   Binary relations:
in the genome   Positional correlations   ← Genome comparisons
                   Gene–gene (indirect) interactions   ← DNA chips
                   Protein–protein (direct) interactions ← Protein chips
                   Substrate–product relations   ← Biochemical knowledge
                   Hierarchial relations   ← Sequence analysis

Fig. 4.13. A strategy for network reconstruction from genomic information.

product) is searched for in the reference network according to the orthologous relation. The orthologues are identified by sequence similarity supplemented by the
positional information of genes in the genome. Third, an *ab initio* type prediction is
performed, based on path computation from a set of binary relations. This is the most
challenging part of the reconstruction process where different types of data and
knowledge are integrated in the form of binary relations and are used to expand our
current knowledge of biochemical networks. As we have seen, path computation of
chemical reactions suggests possible connections of enzymes. The clustering of co-
expressed genes in the operon structures or in the microarray gene expression profiles suggests functional correlations in the biochemical network. The analysis of
genome sequences upstream of the co-expressed genes reveals common sequence
features, which can be used for additional grouping of genes. With the arrival of new
experimental technologies, direct protein–protein interactions may be detected
more systematically. In essence, path computation is logical reasoning. The computerized path computation for a set of binary relations is an automation of human reasoning steps for efficient handling of massive amounts of data and knowledge.

Gene regulatory network

The concept of a gene regulatory network is a natural extension of genetic determinism. An ordered network of specific gene expressions determines, for example,
how an embryo develops or how a cell responds to external stimuli. The genome
contains not only the templates of genes but also the regulatory signals that ultimately determine the network of gene expressions. Of course, the network is not

really deterministic; it depends on the initial conditions of the germ cell and also on many environmental factors. However, the genome is still a blueprint of life—it contains instructions, or programs, on how to respond to the environment.

In contrast, our view presented in Fig. 4.12 may appear to be just the opposite. Gene expression is a mode of protein–protein interaction, which is used by the cell to limit and select the number of proteins in stock. The genome is a data storage of master copies of proteins (and some RNAs too) in compressed forms and the so-called regulatory signals are simply bar codes to retrieve them. When, where, and how retrieval is made is determined by the distributed (non-centralized) network of interacting molecules in the cell, given the initial network of the germ cell and dynamically changing environmental conditions. Thus, the network contains a chemical blueprint of life—a logic of network evolution (see *Genetic and chemical blueprints of life* in Chapter 1)

These two contrasting views of the genome being either the program or the data are not necessarily exclusive. Remember that the program and the data are separate entities in the traditional programming languages, but they are unified in logic programming—rules and facts are just two different forms of predicates (see *Deductive databases* in Section 2.2). The data representation of binary relations, in our case, also contains program specification (logic) of how they are utilized for computation (deduction). When the genome is viewed from this perspective, it is both the program and the data. For example, the information of genes in the genome represents biological logic of gene–gene interactions which are indirect, long-range protein–protein interactions. Thus, in terms of binary relations the gene regulatory network and the protein–protein interaction network can be integrated. It is unlikely that either of the two blueprints, chemical or genetic, is sufficient for understanding life. Our approach based on binary relations provides a practical way to integrate the blueprints of both the chemical logic of network evolution and the genetic logic of biological constraints (Table 4.4).

Network principles

Complete genome sequences give us new opportunities to examine how much network information is actually encoded in the genome. While many key genes are highly conserved among species, comparative genomics has revealed the presence of a fair portion of genes that are apparently unique to each species. This is similar to the architecture of the metabolic network consisting of conserved intermediary

Table 4.4. Genetic and chemical blueprints of life

Blueprint	Entity	Information	
Genetic blueprint of life	Genome	Centralized	Static
Chemical blueprint of life	Network of interacting molecules in the cell	Distributed	Dynamic

metabolism and divergent secondary metabolism. Life requires a set of genes for basic cell processes and cell architecture, but it also requires an additional set of genes for specific interactions with the environment.

Figure 4.14 illustrates what we think are two general principles of biochemical networks encoded in the genome: hierarchy and duality. The duality of chemical logic and genetic logic has just been mentioned (Fig. 4.14(d)). It also exists in the metabolic network which is both a chemical network of compounds and an enzyme network of gene products (Fig. 4.14(c)). An example of hierarchy (Fig. 4.14(a)) is the distinction between a general transcription factor and a specific transcription factor in the gene regulatory network. A general transcription factor, such as the sigma factor in bacteria, can change the transcription activity of many genes at the same time depending on whether the cell is in a normal condition or a stressful condition.

Hierarchy — conservation and diversification

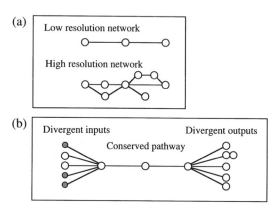

Duality — chemical logic and genetic logic

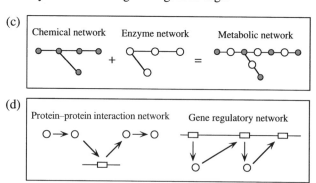

Fig. 4.14 Principles of the biochemical network encoded in the genome.

In the metabolic network we have already seen a similar type of hierarchy, in that a selected number of intermediary compounds serve as hubs in the network. The hierarchical architecture for divergent inputs and conserved processing (Fig. 4.14(b)) is not limited to metabolism. The architecture of signal transduction pathways is similarly organized. Divergent signals are received by many cell surface receptors but they are processed by the conserved pathway, such as the MAP kinase signalling pathway or the second messenger signalling pathway. Because the outputs are also divergent transcription factors to control many genes, the conserved signalling pathway probably represents a low resolution network. At the moment, we lack the knowledge of interacting partners and scaffolds that might account for specific input–output relations in a high resolution network.

Although in principle the chemical and genetic blueprints of life can be integrated, there still remains a conceptual gap. Because the genome is a node, or a reactant, in the network, it should be viewed as a dynamic entity that can be modified. This may not be a problem for the genetic blueprint of life. Although the genome is known to undergo, for example, methylations, recombinations, and repeat expansions, these are minor modifications that modulate the gene regulatory network during information expression. The genome is still assumed to be a static entity for information transmission of the repertoire of genes. In contrast, according to the chemical blueprint of life, the genome is just one of the molecules in the network of molecular interactions. The repertoire of genes itself may undergo more frequent modifications.

An important observation by comparative genomics is the extensive rearrangement of orthologous genes, even for closely related species. In addition, each species contains many groups of paralogous genes of varying sizes, which also appear to be extensively shuffled. There have been attempts to define genomic transformation, or to find a minimum number of steps required for conversion of the gene order from one genome to another, but it remains to be seen whether there is any underlying principle of genome rearrangement. Furthermore, it has been argued that horizontal gene transfer across species must be commonly occurring to account for the presence of many genes with foreign characteristics. These observations suggest that the genome is a dynamic entity, at least, at the level of speciation.

Complex systems

It is not inconceivable that our view of the genome is biased by past biological experiments that used stable laboratory strains which were easy to culture, but which are no way representative of all biological species. Life is, in essence, an open system. The very idea of an organism's stability is a reflection of its isolation from a dynamic environment. Again by analogy to protein folding, the view of spontaneous 3D structure formation was first established by thermodynamic experiments with small stable proteins, such as ribonuclease and lysozyme. This view had to be modified later, and protein folding *in vivo* is now considered part of the dynamic process of interacting molecules. As we learn more from the complete genome sequences that reflect a better sampling of all biological species, we might find the genome to be a far more dynamic entity than we presently assume.

The conceptual framework of binary relation and deduction (path computation), as well as its practical implementation of KEGG, may appear to be applicable only to the static aspects of the genome and the biochemical network. In contrast, there have been attempts to simulate time-dependent behaviours of metabolic pathways, for example, by a set of differential equations or by a type of graph called a Petri net. Our formalism of the KEGG system is not intended for this type of simulation, because a simple collection of what is already known is unlikely to provide any new biological insights. The collective behaviour of the biological network results from complex non-linear interactions within the network and with the dynamic environment, about which our knowledge is largely incomplete. It is thus essential to uncover such interactions and to understand general principles of the network behaviour. KEGG is a library of virtual cells—computer representations of the known biochemical networks in living cells of different organisms—that can be used to simulate and predict dynamic responses to perturbations (see Fig. 1.14). A general protocol of simulation is to supplement the known network with a computationally generated set of binary relations. The collective behaviour of the simulated network should predict and explain actual experimental observations in living cells, such as dynamic changes of gene expression profiles in response to given perturbations.

At the moment, we limit our exploration of post-genome informatics at the molecular network level toward understanding the behaviour of a single cell from the network of interacting molecules and genes. This is actually one example of the so-called complex systems (Table 4.5). The study of complex systems spans many different disciplines at different levels of abstraction. The complex systems share a common architecture consisting of many interacting components whose collective behaviour becomes quite complex, even though the behaviours of individual components may be simple. This is because of non-linear interactions between components and with a dynamic environment. Eventually, still higher levels of self-organizations in the biological complex systems, such as brain function, ecosystem, and even human civilization, may be connected to the information of the genome—thus becoming the subjects of post-genome informatics.

Table 4.5. Biological examples of complex systems

System	Node	Edge
Protein 3D structure	Atom	Atomic interaction
Organism	Molecule	Molecular interaction
Brain	Cell	Cellular interaction
Ecosystem	Organism	Organism interaction
Civilization	Human	Human interaction

Appendix. Methods in computational molecular biology—Bibliography

1. Sequence analysis I. Sequence alignment

1.1 *Pairwise sequence alignment*

Historic papers on dynamic programming

Needleman, S. B. and Wunsch, C. D. (1970). A general method applicable to the search for similarities in the amino acid sequence of two proteins. *J. Mol. Biol.*, **48**, 443–53.

Sankoff, D. (1972). Matching sequences under deletion/insertion constraints. *Proc. Natl. Acad. Sci. USA*, **69**, 4–6.

Local alignment

Sellers, P. H. (1980). The theory and computation of evolutionary distances: pattern recognition. *J. Algorithms*, **1**, 359–73.

Smith, T. F. and Waterman, M. S. (1981). Identification of common molecular subsequences. *J. Mol. Biol.*, **147**, 195–7.

Goad, W. B. and Kanehisa, M. I. (1982). Pattern recognition in nucleic acid sequences. I. A general method for finding local homologies and symmetries. *Nucleic Acids Res.*, **10**, 247–63.

Similarity matrices

Dayhoff, M. O., Schwartz, R. M., and Orcutt, B. C. (1978). A model of evolutionary change in proteins. In *Atlas of protein sequence and structure*, Vol. 5, Suppl. 3 (ed. M. O. Dayhoff), pp. 345–52. National Biomedical Research Foundation, Washington, DC.

Henikoff, S. and Henikoff, J. G. (1992). Amino acid substitution matrices from protein blocks. *Proc. Natl. Acad. Sci. USA*, **89**, 10915–19.

Gap penalties

Gotoh, O. (1982). An improved algorithm for matching biological sequences. *J. Mol. Biol.*, **162**, 705–8.

Kanehisa, M. I. and Goad, W. B. (1982). Pattern recognition in nucleic acid sequences. II. An efficient method for finding locally stable secondary structures. *Nucleic Acids Res.*, **10**, 265–78.

Diagonals

Ukkonen, E. (1983). On approximate string matching. *Lecture Notes Comp. Sci.*, **158**, 487–95.

Fickett, J. W. (1984). Fast optimal alignment. *Nucleic Acids Res.*, **12**, 175–9.

1.2   Database search

FASTA algorithm

Wilbur, W. J. and Lipman, D. J. (1983). Rapid similarity searches of nucleic acid and protein data banks. *Proc. Natl. Acad. Sci. USA*, **80**, 726–30.

Lipman, D. J. and Pearson, W. R. (1985). Rapid and sensitive protein similarity searches. *Science*, **227**, 1435–41.

BLAST algorithm

Altschul, S. F., Gish, W., Miller, W., Myers, E. W., and Lipman, D. J. (1990). Basic local alignment search tool. *J. Mol. Biol.*, **215**, 403–10.

Altschul, S. F., Madden, T. L., Schaeffer, A. A., Zhang, J., Zhang, Z., Miller, W., and Lipman, D. J. (1997). Gapped BLAST and PSI-BLAST: a new generation of protein database search programs. *Nucleic Acids Res.*, **25**, 3389–402.

Statistical significance

Karlin, S. and Altschul, S. F. (1990). Methods for assessing the statistical significance of molecular sequence features by using general scoring schemes. *Proc. Natl. Acad. Sci. USA*, **87**, 2264–8.

Pearson, W. R. (1995). Comparison of methods for searching protein sequence databases. *Protein Sci.*, **4**, 1145–60.

1.3   Multiple sequence alignment

Multi-dimensional dynamic programming

Waterman, M. S., Smith, T. F., and Beyer, W. A. (1976). Some biological sequence metrics. *Adv. Math.*, **20**, 367–87.

Murata, M., Richardson, J. S., and Sussman, J. L. (1985). Simultaneous comparison of three protein sequences. *Proc. Natl. Acad. Sci. USA*, **82**, 3073–7.

Multi-dimensional diagonals

Carrillo, H. and Lipman, D. (1988). The multiple sequence alignment problem in biology. SIAM *J. Appl. Math.*, **48**, 1073–82.

Lipman, D. J., Altschul, S. F., and Kececioglu, J. D. (1989). A tool for multiple sequence alignment. *Proc. Natl. Acad. Sci. USA*, **86**, 4412–15.

Progressive multiple alignment

Feng, D.-F. and Doolittle, R. F. (1987). Progressive sequence alignment as a prerequisite to correct phylogenetic trees. *J. Mol. Evol.*, **25**, 351–60.

Higgins, D. G., Bleasby, A. J., and Fuchs, R. (1992). CLUSTAL V: improved software for multiple sequence alignment. *Comput. Appl. Biosci.*, **8**, 189–91.

Thompson, J. D., Higgins, D. G., and Gibson, T. J. (1994). CLUSTAL W: improving the sensitivity of progressive multiple sequence alignment through sequence weighting, position-specific gap penalties and weight matrix choice. *Nucleic Acids Res.*, **22**, 4673–80.

Iterative multiple alignment

Berger, M. P. and Munson, P. J. (1991). A novel randomized iterative strategy for aligning multiple protein sequences. *Comput. Appl. Biosci.*, **7**, 479–84.

Gotoh, O. (1993). Optimal alignment between groups of sequences and its application to multiple sequence alignment. *Comput. Appl. Biosci.*, **9**, 361–70.

Simulated annealing

Ishikawa, M., Toya, T., Hoshida, M., Nitta, K., Ogiwara, A., and Kanehisa, M. (1993). Multiple sequence alignment by parallel simulated annealing. *Comput. Appl. Biosci.*, **9**, 267–73.

Phylogenetic analysis

Saitou, N. and Nei, M. (1987). The neighbor-joining method: a new method for reconstructing phylogenetic trees. *J. Mol. Evol.*, **4**, 406–25.

Sneath, P. H. A. and Sokal, R. R. (1973). *Numerical taxonomy.* Freeman, San Francisco.

1.4 RNA secondary structure prediction

Historic papers on RNA secondary structure prediction

Tinoco, I., Jr, Uhlenbeck, O. C., and Levine, M. D. (1971). Estimation of secondary structure in ribonucleic acids. *Nature*, **230**, 362–7.

Pipas, J. M. and McMahon, J. E. (1975). Method for predicting RNA secondary structure. *Proc. Natl. Acad. Sci. USA*, **72**, 2017–21.

Studnicka, G. M., Rahn, G. M., Cummings, I. W., and Salser, W. A. (1978). Computer method for predicting the secondary structure of single-stranded RNA. *Nucleic Acids Res.*, **5**, 3365–87.

Dynamic programming

Nussinov, R. and Jacobson, A. B. (1980). Fast algorithm for predicting the secondary structure of single-stranded RNA. *Proc. Natl. Acad. Sci. USA*, **77**, 6309–13.

Zuker, M. and Stiegler, P. (1981). Optimal computer folding of large RNA sequences using thermodynamics and auxiliary information. *Nucleic Acids Res.*, **9**, 133–48.

Zuker, M. (1989). On finding all suboptimal foldings of an RNA molecule. *Science*, **244**, 48–52.

Free energy values

Salser, W. (1977). Globin mRNA sequences: Analysis of base pairing and evolutionary implications. *Cold Spring Harb. Symp. Quant. Biol.*, **42**, 985–1002.
Turner, D. H., Sugimoto, N., Jaeger, J. A., Longfellow, C. E., Freier, S. M., and Kierzek, R. (1987). Improved parameters for prediction of RNA structure. *Cold Spring Harb. Symp. Quant. Biol.*, **52**, 123–33.

Hopfield neural network

Hopfield, J. J. and Tank, D. W. (1986). Computing with neural circuits: a model. *Science*, **233**, 625–33.

Formal grammar

Searls, D. B. (1993). The computational linguistics of biological sequences. In *Artificial intelligence and molecular biology* (ed. L. Hunter), pp. 47–120. AAAI Press/MIT Press.
Sakakibara, Y., Brown, M., Hughey, R., Mian, I. S., Sjolander, K., Underwood, R. C., and Haussler, D. (1994). Stochastic context-free grammars for tRNA modeling. *Nucleic Acids Res.*, **22**, 5112–20.

2.   Sequence analysis II. Sequence features

2.1   *Protein secondary structure prediction*

Historic papers on protein secondary structure prediction

Ptitsyn, O. B. and Finkelstein, A. V. (1983). Theory of protein secondary structure and algorithm of its prediction. *Biopolymers*, **22**, 15–25.
Chou, P. Y. and Fasman, G. D. (1974). Prediction of protein conformation. *Biochemistry*, **13**, 222–44.
Garnier, J., Osguthorpe, D. J., and Robson, B. (1978). Analysis of the accuracy and implications of simple methods for predicting the secondary structure of globular proteins. *J. Mol. Biol.*, **120**, 97–120.
Lim, V. I. (1974). Algorithms for prediction of α-helical and β-structural regions in globular proteins. *J. Mol. Biol.*, **88**, 873–94.
Cohen, F. E., Abarbanel, R. M., Kuntz, I. D., and Fletterick, R. J. (1986). Turn prediction in proteins using a pattern-matching approach. *Biochemistry*, **25**, 266–75.

Neural network

Qian, N. and Sejnowski, T. J. (1988). Predicting the secondary structure of globular proteins using neural network models. *J. Mol. Biol.*, **202**, 865–84.
Rost, B. and Sander, C. (1993). Improved prediction of protein secondary structure by use of sequence profiles and neural networks. *Proc. Natl. Acad. Sci. USA*, **90**, 7558–62.

Hidden Markov model

Asai, K., Hayamizu, S., and Handa, K. (1993). Prediction of protein secondary structure by the hidden Markov model. *Comput. Appl. Biosci.*, **9**, 141–6.

Logic programming

Muggleton, S., King, R. D., and Sternberg, M. J. (1992). Protein secondary structure prediction using logic-based machine learning. *Protein Eng.*, **5**, 647–57.

Definition of secondary structures

Kabsch, W. and Sander, C. (1983). Dictionary of secondary structure: pattern recognition of hydrogen-bonded and geometric features. *Biopolymers*, **22**, 2577–637.

2.2   Protein families and sequence motifs

Motifs

Smith, H. O., Annau, T. M., and Chandrasegaran, S. (1990). Finding sequence motifs in groups of functionally related proteins. *Proc. Natl. Acad. Sci. USA*, **87**, 826–30.

Henikoff, S. and Henikoff, J. G. (1991). Automated assembly of protein blocks for database searching. *Nucleic Acids Res.*, **19**, 6565–72.

Ogiwara, A., Uchiyama, I., Seto, Y., and Kanehisa, M. (1992). Construction of a dictionary of sequence motifs that characterize groups of related proteins. *Protein Eng.*, **5**, 479–88.

Profiles

Gribskov, M., McLachlan, A. D., and Eisenberg, D. (1987). Profile analysis: detection of distantly related proteins. *Proc. Natl. Acad. Sci. USA*, **84**, 4355–8.

Ogiwara, A., Uchiyama, I., Takagi, T., and Kanehisa, M. (1996). Construction and analysis of a profile library characterizing groups of structurally known proteins. *Protein Sci.*, **5**, 1991–9.

Hidden Markov model

Sonnhammer, E. L., Eddy, S. R., and Durbin, R. (1997). Pfam: a comprehensive database of protein domain families based on seed alignments. *Proteins*, **28**, 405–20.

Yada, T., Totoki, Y., Ishikawa, M., Asai, K., and Nakai, K. (1998). Automatic extraction of motifs represented in the hidden Markov model from a number of DNA sequences. *Bioinformatics*, **14**, 317–25.

Statistics

Lawrence, C. E. and Reilly, A. A. (1990). An expectation maximization (EM) algorithm for the identification and characterization of common sites in unaligned biopolymer sequences. *Proteins*, **7**, 41–51.

Lawrence, C. E., Altschul, S. F., Boguski, M. S., Liu, J. S., Neuwald, A. F., and Wootton, J. C. (1993). Detecting subtle sequence signals: a Gibbs sampling strategy for multiple alignment. *Science*, **262**, 208–14.

Information theory

Stormo, G. D. and Hartzell, G. W. III (1989). Identifying protein-binding sites from unaligned DNA fragments. *Proc. Natl. Acad. Sci. USA*, **86**, 1183–7.

Kohonen neural network

Kohonen, T. (1982). Self-organized formation of topologically correct feature maps. *Biol. Cybern.*, **43**, 59–69.

Ferran, E. A., Pflugfelder, B., and Ferrara, P. (1994). Self-organized neural maps of human protein sequences. *Protein Sci.*, **3**, 507–21.

2.3   Functional predictions

Neural network

Stormo, G. D., Schneider, T. D., Gold, L., and Ehrenfeucht, A. (1982). Use of the 'perceptron' algorithm to distinguish translational initiation sites in *E. coli. Nucleic Acids Res.*, **10**, 2997–3011.

Nakata, K., Kanehisa, M., and DeLisi, C. (1985). Prediction of splice junctions in mRNA sequences. *Nucleic Acids Res.*, **13**, 5327–40.

Horton, P. B. and Kanehisa, M. (1992). An assessment of neural network and statistical approaches for prediction of *E. coli* promoter sites. *Nucleic Acids Res.*, **20**, 4331–8.

Hidden Markov model

Krogh, A., Brown, M., Mian, I. S., Sjolander, K., and Haussler, D. (1994). Hidden Markov models in computational biology. Applications to protein modeling. *J. Mol. Biol.*, **235**, 1501–31.

Brown, M., Hughey, R., Krogh, A., Mian, I. S., Sjolander, K., and Haussler, D. (1993). Using dirichlet mixture priors to derive hidden Markov models for protein families. *ISMB*, **1**, 47–55.

Discriminant analysis

Kyte, J. and Doolittle, R. F. (1982). A simple method for displaying the hydropathic character of a protein. *J. Mol. Biol.*, **157**, 105–32.

Klein, P., Kanehisa, M., and DeLisi, C. (1985). The detection and classification of membrane-spanning proteins. *Biochim. Biophys. Acta*, **815**, 468–76.

von Heijne, G. (1992). Membrane protein structure prediction. Hydrophobicity analysis and the positive-inside rule. *J. Mol. Biol.*, **225**, 487–94.

Kihara, D., Shimizu, T., and Kanehisa, M. (1998). Prediction of membrane proteins based on classification of transmembrane segments. *Protein Eng.*, **11**, 961–70.

Chou, K.-C. and Elrod, D. W. (1999). Protein subcellular location prediction. *Protein Eng.*, **12**, 107–18.

Expert system

Nakai, K. and Kanehisa, M. (1991). Expert system for predicting protein localization sites in Gram-negative bacteria. *Proteins*, **11**, 95–110.

Nakai, K. and Kanehisa, M. (1992). A knowledge base for predicting protein localization sites in eukaryotic cells. *Genomics*, **14**, 897–911.

Cluster analysis

Nakai, K., Kidera, A., and Kanehisa, M. (1988). Cluster analysis of amino acid indices for prediction of protein structure and function. *Protein Eng.*, **2**, 93–100.

Tomii, K. and Kanehisa, M. (1996). Analysis of amino acid indices and mutation matrices for sequence comparison and structure prediction of proteins. *Protein Eng.*, **9**, 27–36.

2.4   Gene finding

Neural network

Uberbacher, E. C. and Mural, R. J. (1991). Locating protein-coding regions in human DNA sequences by a multiple sensor-neural network approach. *Proc. Natl. Acad. Sci. USA*, **88**, 11261–5.

Hidden Markov model

Burge, C. and Karlin, S. (1997). Prediction of complete gene structures in human genomic DNA. *J. Mol. Biol.*, **268**, 78–94.

Markov chain

Borodovsky, M. and McIninch, J. (1993). GENMARK: Parallel gene recognition for both DNA strands. *Computers Chem.*, **17**, 123–33.

3.   Structure analysis

3.1   Protein structure comparison

3D alignment

Kabsch, W. (1976). A solution for the best rotation to relate two sets of vectors. *Acta Cryst.*, **A32**, 922–3.

Rossmann, M. G. and Argos, P. (1976). Exploring structural homology of proteins. *J. Mol. Biol.*, **105**, 75–95.

Remington, S. J. and Matthews, B. W. (1980). A systematic approach to the comparison of protein structures. *J. Mol. Biol.*, **140**, 77–99.

Taylor, W. R. and Orengo, C. A. (1989). Protein structure alignment. *J. Mol. Biol.*, **208**, 1–22.

Sali, A. and Blundell, T. L. (1990). Definition of general topological equivalence in protein structures: a procedure involving comparison of properties and relationships through simulated annealing and dynamic programming. *J. Mol. Biol.*, **212**, 403–28.

2D or 1D alignment

Holm, L. and Sander, C. (1993). Protein structure comparison by alignment of distance matrices. *J. Mol. Biol.*, **233**, 123–38.

Matsuo, Y. and Kanehisa, M. (1993). An approach to systematic detection of protein structural motifs. *Comput. Appl. Biosci.*, **9**, 153–9.

Fold classification

Levitt, M. and Chothia, C. (1976). Structural patterns in globular proteins. *Nature*, **261**, 552–8.

Orengo, C. A., Flores, T. P., Taylor, W. R., and Thornton, J. M. (1993). Identification and classification of protein fold families. *Protein Eng.*, **6**, 485–500.

Murzin, A. G., Brenner, S. E., Hubbard, T., and Chothia, C. (1995). SCOP: a structural classification of proteins database for the investigation of sequences and structures. *J. Mol. Biol.*, **247**, 536–40.

Mizuguchi, K. and Go, N. (1995). Comparison of spatial arrangements of secondary structural elements in proteins. *Protein Eng.*, **8**, 353–62.

3.2   Protein 3D structure prediction

Comparative modelling

Blundell, T. L., Sibanda, B. L., Sternberg, M. J., and Thornton, J. M. (1987). Knowledge-based prediction of protein structures and the design of novel molecules. *Nature*, **326**, 347–52.

Sander, C. and Schneider, R. (1991). Database of homology-derived protein structures and the structural meaning of sequence alignment. *Proteins*, **9**, 56–68.

Sali, A. and Blundell, T. L. (1993). Comparative protein modelling by satisfaction of spatial restraints. *J. Mol. Biol.*, **234**, 779–815.

Threading

Bowie, J. U., Luthy, R., and Eisenberg, D. (1991). A method to identify protein sequences that fold into a known three-dimensional structure. *Science*, **253**, 164–70.

Jones, D. T., Taylor, W. R., and Thornton, J. M. (1992). A new approach to protein fold recognition. *Nature*, **358**, 86–9.

Sippl, M. J. and Weitckus, S. (1992). Detection of native-like models for amino acid sequences of unknown three-dimensional structure in a database of known protein conformations. *Proteins*, **13**, 258–71.

Nishikawa, K. and Matsuo, Y. (1993). Development of pseudoenergy potentials for assessing protein 3D–1D compatibility and detecting weak homologies. *Protein Eng.*, **6**, 811–20.

Rost, B., Schneider, R., and Sander, C. (1997). Protein fold recognition by prediction-based threading. *J. Mol. Biol.*, **270**, 471–80.

Ab initio modelling

Srinivasan, R. and Rose, G. D. (1995). LINUS: A hierarchic procedure to predict the fold of a protein. *Proteins*, **22**, 81–99.

Skolnick, J., Kolinski, A., and Ortiz, A. R. (1997). MONSSTER: a method for folding globular proteins with a small number of distance restraints. *J. Mol. Biol.*, **265**, 217–41.

Dandekar, T. and Argos, P. (1997). Applying experimental data to protein fold prediction with the genetic algorithm. *Protein Eng.*, **10**, 877–93.

Docking

Kuntz, I. D., Blaney, J. M., Oatley, S. J., Langridge, R., and T. E. Ferrin (1982). A geometric approach to macromolecule–ligand interactions. *J. Mol. Biol.*, **161**, 269–88.

Gabb, H. A., Jackson, R. M., and Sternberg, M. J. (1997). Modelling protein docking using shape complementarity, electrostatistics and biochemical information. *J. Mol. Biol.*, **272**, 106–20.

3.3 RNA 3D structure modelling

Distance geometry

Hubbard, J. M. and Hearst, J. E. (1991). Predicting the three-dimensional folding of transfer RNA with a computer modeling protocol. *Biochemistry*, **30**, 5458–65.

Constraint satisfaction

Major, F., Turcotte, M., Gautheret, D., Lapalme, G., Fillion, E., and Cedergren, R. (1991). The combination of symbolic and numerical computation for three-dimensional modeling of RNA. *Science*, **253**, 1255–60.

Genetic algorithm

Ogata, H., Akiyama, Y., and Kanehisa, M. (1995). A genetic algorithm based molecular modeling technique for RNA stem-loop structures. *Nucleic Acids Res.*, **23**, 419–26.

4.   Network analysis

4.1   Genome analysis

Gene clusters and orthologues

Mushegian, A. R. and Koonin, E. V. (1996). A minimal gene set for cellular life derived by comparison of complete bacterial genomes. *Proc. Natl. Acad. Sci. USA*, **93**, 10268–73.

Tatusov, R., Koonin, E. V., and Lipman, D. J. (1997). A genomic perspective on protein families. *Science*, **278**, 631–7.

Tomii, K. and Kanehisa, M. (1998). A comparative analysis of ABC transporters in complete microbial genomes. *Genome Res.*, **8**, 1048–59.

Dandekar, T., Snel, B., Huynen, M., and Bork, P. (1998). Conservation of gene order: a fingerprint of proteins that physically interact. *Trends Biochem. Sci.*, **23**, 324–8.

Overbeek, R., Fonstein, M., D'Souza, M., Pusch, G. D., and Maltsev, N. (1999). The use of gene clusters to infer functional coupling. *Proc. Natl. Acad. Sci. USA*, **96**, 2896–901.

Expression analysis

DeRisi, J. L., Iyer, V. R., and Brown, P. O. (1997). Exploring the metabolic and genetic control of gene expression on a genomic scale. *Science*, **278**, 680–6.

Eisen, M. B., Spellman, P. T., Brown, P. O., and Botstein, D. (1998). Cluster analysis and display of genome-wide expression patterns. *Proc. Natl. Acad. Sci. USA*, **95**, 14863–8.

4.2   Pathway analysis

Pathway database

Kanehisa, M. (1997). A database for post-genome analysis. *Trends Genet.*, **13**, 375–6.

Kanehisa, M. (1997). Linking databases and organisms: GenomeNet resources in Japan. *Trends Biochem. Sci.*, **22**, 442–4.

Goto, S., Nishioka, T., and Kanehisa, M. (1998). LIGAND: chemical database for enzyme reactions. *Bioinformatics*, **14**, 591–9.

Ogata, H., Goto, S., Sato, K., Fujibuchi, W., Bono, H., and Kanehisa, M. (1999). KEGG: Kyoto Encyclopedia of Genes and Genomes. *Nucleic Acids Res.*, **27**, 29–34.

Network comparison

Ogata, H., Fujibuchi, W., Goto, S., and Kanehisa, M. (2000). A heuristic graph comparison algorithm and its application to detect functionally related enzyme clusters. *Nucleic Acids Res.* **28**, 4021–8.

Fujibuchi, W., Ogata, H., Matsuda, H., and Kanehisa, M. (2000). Automatic detection of conserved gene clusters in multiple genomes by graph comparison and P-quasi grouping. *Nucleic Acids Res.* **28**, 4029–36.